How to Build a Cyber-Resilient Organization

Internal Audit and IT Audit

Series Editor:

Dan Swanson, Dan Swanson and Associates Ltd, Winnipeg, Manitoba, Canada

The *Internal Audit and IT Audit* series publishes leading-edge books on critical subjects facing audit executives as well as internal and IT audit practitioners. Key topics include Audit Leadership, Cybersecurity, Strategic Risk Management, Auditing Various IT Activities and Processes, Audit Management, and Operational Auditing.

For more information about this series, please visit https://www.crcpress.com/Internal-Audit-and-IT-Audit/book-series/CRCINTAUDITA

How to Build a Cyber-Resilient Organization

Dan Shoemaker, Anne Kohnke, and Ken Sigler

CRC Press
Taylor & Francis Group
Boca Raton London New York

CRC Press is an imprint of the
Taylor & Francis Group, an **informa** business
AN AUERBACH BOOK

CRC Press
Taylor & Francis Group
6000 Broken Sound Parkway NW, Suite 300
Boca Raton, FL 33487-2742

International Standard Book Number-13: 978-1-138-55819-9 (Paperback)

Visit the Taylor & Francis Web site at
http://www.taylorandfrancis.com

and the CRC Press Web site at
http://www.crcpress.com

Contents

Foreword

The goal of a cybersecurity strategy is to prevent the compromise of resources. Attackers attempt to compromise networks, computers, and other resources by finding weak points in the cybersecurity policy, mechanisms, or strategy. The points at which they can reach the target system is called the *attack surface*. Cybersecurity policies and mechanisms aim to secure this surface.

Cyber resilience, as defined in this book, is a strategy to minimize this surface, and so make it less onerous to protect. It asks what the enterprise *needs* to protect to continue functioning successfully. Once these critical assets are identified, the architecture of the system is developed in such a way as to ensure that defenses protecting these assets are in place and effective. The rest of the enterprise can then be secured. The advantage of this asset-based approach is that the enterprise knows what assets it needs to protect, to what degree the protection must succeed, and what will happen should those protections fail.

It also aids in developing recovery mechanisms should those protections fail. In its traditional sense, "resilience" implies that the system may degrade, but it will recover. Here, given the knowledge of the critical assets, the enterprise can take steps to ensure that the system architecture keeps at least some of the asset available and functioning, even if not at the peak level, until the compromise is dealt with. Business functionality may be degraded for a time, but it will recover.

Developing a cyber-resilient plan requires both risk analysis and planning. First, the risk analysis shows what will happen when assets become compromised or fail. The enterprise and its stakeholders then determine what is acceptable to them—and, more importantly, what is not. This identifies the critical business functionality of the enterprise, and from that the critical assets can be identified. The cost of failure can be prioritized, giving an ordering of the critical assets. This enables the enterprise to deploy its protective resources to its greatest benefit.

The idea of beginning with the critical assets runs counter to the way much cybersecurity is done. Generally, the attack surface of the system is defined by examining possibly the entire system, locating points of potential weaknesses, and monitoring or strengthening them. The problem with using *only* this approach is twofold. First, the attack surface defined in this way is very broad. Second, the attack surface covers both critical and noncritical assets *without distinction*. This last point is particularly important, as the enterprise might protect something very well that, when it is compromised, does little to no damage to the business of the enterprise, and not protect something very well that, when it is compromised, disrupts the business function of the enterprise. Its lack of discrimination of criticality may result in the enterprise allocating its protection resources inappropriately.

Perhaps even more important is the need to locate the assets, understand them and how they play into the mission of the enterprise, and what the cybersecurity policy relevant to that asset should be. All too often, institutions lose track of assets or fail to understand how the functionality of one asset affects another asset. The risk analysis and prioritization needed to develop a cyber resilience strategy as discussed in this book will, if done correctly, ensure that the enterprise knows what its critical assets are, how they affect one another, what the requirements of the policy are in order to "secure the asset," and how that policy should be implemented.

This book presents a methodology for developing a cyber resilience strategy. It will aid enterprises in securing their resources. Thus, it fills a need, and fills it very well.

Matt Bishop
Davis, CA
July 12, 2018

Preface

You Can't Secure Everything

Persistent threats are so all-encompassing in the virtual ecosystem that it is unrealistic to think that you can defend against all of them. This means that successful cyberattacks will occur, no matter how much money we throw at the problem. So, organizations need to adopt a strategy that will ensure that its critical assets are protected, no matter what other items are compromised. The aim of this text is to present a novel new approach that has been devised to do that very thing.

At present, we secure systems at all logical points of access. Even defense-in-depth schemes simply involve the creation of increasingly rigorous perimeters. This new approach makes the practical assumption that instead of defending assets at all points of access, the organization ensures that only the things that the organization simply can't live without are unconditionally protected.

In concept, this strategy would narrow the protection scheme to the point where absolute assurance could be guaranteed for only some items while reducing the impact and recovery time of the rest of the organization's noncritical assets. This selective versus universal approach to security has been termed "cyber resilience."

Why Cyber Resilience?

Cyber resilience has a much narrower focus than cybersecurity. At its heart, cybersecurity concentrates a targeted set of protection measures on every viable point of access. Because there are a lot of potential access points, a cybersecurity solution requires a very large resource commitment.

Cyber resilience focuses on identifying and locking down just those assets that the organization needs to ensure the survival of the business. Then, cyber resilience conducts a disciplined recovery process that ensures the timely restoration of its noncritical assets to an acceptable level of operation, with the minimum amount of harm.

The concept of cyber resilience is supported by three of Saltzer and Schroeder's lesser-known principles.

1. *Economy of mechanism*: Keep the design as simple and small as possible.
2. *Least common mechanism*: Minimize the amount of mechanism common to all users.
3. *Work factor*: The cost must be greater than the potential attacker is willing to commit.

These principles make the cyber resilience concept more effective and less resource intensive than other approaches. Economy of mechanism locks up just the critical assets. This allows for simplicity in the design and implementation of the protection. It also concentrates resources on ensuring the protection of the critical assets rather than diffusing the investment across all assets. Most importantly, if the protection of the critical asset is made robust enough, attacking it will become too expensive and time consuming for the attacker, forcing them to move to more vulnerable targets.

Purpose of the Text

You will learn how to create a verifiable cyber-resilient infrastructure, which will ensure reliable security for critical objects. The book will explain how to establish systematic identification, prioritization, protection, and recovery processes. This is embodied in seven generic principles:

1. *Classify*: You can't protect things that you don't know exist. Thus, all the organization's assets must be identified, labeled, and arrayed in a coherent baseline of "things."
2. *Risk*: Resiliency requires appropriate situational awareness. Therefore, all known threat scenarios as they apply to the identified asset base must be identified and evaluated.
3. *Rank*: The assets that the organization absolutely can't afford to lose are prioritized, and provably effective countermeasures are designed for each of the priority assets.
4. *Architecture*: Resilience is baked into the architecture through a targeted set of well-designed and practical controls.
5. *Test*: Architectural resilience is evaluated against stated mission goals.
6. *Recover*: Well-defined processes are established to ensure that all noncritical assets are fully restored within reasonable and effective parameters.
7. *Evolve*: The organization dynamically adjusts its cyber-resilient architecture based on lessons learned over time.

Organization of the Text

This book encapsulates the belief that the creation of a cyber-resilient architecture is a strategic exercise. The outcome of this exercise is a formally defined and implemented infrastructure of best practices specifically aimed at optimizing the survival of critical organizational functions across the organization. As with any complex process, deployment can only be substantiated through a rational and explicit framework of auditable controls. The process for creating and deploying those controls is what is presented in these chapters.

Chapter 1: It's Time for a New Paradigm

The book will detail the general process for creating a cyber-resilient organization. The goal of this chapter is to give the reader an understanding of the overall strategic concept of cyber resilience as well as provide the justification and advantages of cyber resilience as a practical method for assuring organizational survival.

Readers will see how fundamental strategic activities can provide the basis for the absolute assurance of the critical assets of an organization. Readers will understand the differences between cybersecurity and cyber resilience. They will see that the actions taken to ensure cyber resilience are more cost effective and likely to assure organizational survival. Finally, we will describe the strategic planning process for establishing cyber resilience in any organization.

The reader will learn why a formal, comprehensive methodology aimed at ensuring the safety and security of vital assets is critical to the success of a cyber-resilient architecture. The aim of this chapter is to give an overview of the typical process of cyber resilience strategic planning including the necessary associations that must be created between the steps in the process. To ensure cyber resilience, these steps must be fully planned and coordinated across the organization.

Coordination of this degree of complex design and sustainment work requires a common and coherent methodology. The methodology that we outline will give managers practical control over the establishment and operation of a cyber-resilient architecture. The aim of this chapter is to provide readers with the ability to create and operate a cyber-resilient organization.

Chapter 2: Asset Identification and Classification

In some respects, cyber resilience is nothing more than an ultimate defense-in-depth and continuity management solution rolled into a highly focused protection mission. The core aim of cyber resilience is to maintain critical business functionality at all costs. Thus, the decisions that come out of the cyber resilience design process will determine how the business will invest its precious time and resources and build its architecture.

The key to cyber resilience lies in understanding what constitutes core functionality. Cyber resilience assumes that all systems will eventually be compromised. Given this assumption, the cyber resilience function ensures a robust array of specifically targeted controls to ensure that only the subset of functions essential to the continuing operation of the business are fully and completely protected, even if all other system activities are compromised.

In conjunction with the aim of ensuring the survival of core functionality, the cyber resilience process also defines clear, straightforward, and practical paths to restore any lower priority functions that might have been lost in the actual compromise. The seven stages of cyber resilience are designed to achieve those two specific goals. The identification process is perhaps the most important step in creating a cyber-resilient organization. That is because the outcome of the classification process will drive every subsequent protection action.

Logically, if an organization does not understand what assets it has, it is almost impossible to intelligently protect them. Thus, a deliberate, formally executed, and documented classification activity is the key starting point. This is an organization-wide exercise whose aim is to understand the criticality, sensitivity, and priority of all items in the asset base. It involves all stakeholders because buy in is an essential condition for embedding changes in the organization.

Chapter 3: Establishing the Risk Status of the Corporate Infrastructure

Risk assessment provides timely and accurate understanding of the threat status of all components of the asset base, and is essentially a situational awareness function. The risk stage employs situational awareness practices to drive the decisions about the best way to ensure that critical assets and services will continue to function as desired. The aim is to fully understand every hazard in the threat environment that might affect a critical asset. The term "hazard" denotes a threat or an incident, natural or man-made, that warrants action to protect against harm and to minimize disruptions of the mission.

This includes natural disasters, cyber incidents, acts of terrorism, sabotage, and destructive criminal activities targeting critical components of the enterprise infrastructure.

The outcome of this phase is a detailed map of the risk environment sufficient to support decision-making with respect to organizational priorities. No decisions are made until the entire threat assessment is mapped. There is obviously a potential that meaningful threats might be missed or that a new threat might appear after the original risk assessment is completed.

Additionally, there should be a comprehensive plan to ensure subsequent systematic risk assessments against any potential attack that might occur against assets viewed as critical. Since the architecture will be altered to respond to those threats, it is also crucial that a process exists to rapidly respond to those risks. The outcome is reasonable confidence that all conceivable risks to a given asset and its dependencies have been identified, characterized, and ranked for likelihood and impact.

Chapter 4: Prioritization of Assets and Establishing a Plan for Resilient Change

This is the second most important aspect of the cyber resilience process. Once the organization's assets have been identified and baselined and the threat environment characterized, the criticality of all assets in the asset baseline is ranked. This is an organization-wide ranking process involving all stakeholders. "Assets" comprise all the people, processes, technologies, and facilities required to achieve the organizational purpose. However, some assets are more critical than others. The ranking process identifies, documents, and assures only those assets ranked as "critical" to the organization's mission, vision, values, and purposes.

Unfortunately, ranking can often turn into a political free-for-all where various stakeholders attempt to enforce their own agendas. Obviously, this can't be allowed to happen if the eventual architectural solution is going to be truly resilient. Therefore, criticality must be understood based on a clear map of functions and dependencies, which are referenced in an objective and rational way to the mission and goals of the organization. This chapter will discuss how an asset can only be labeled "critical" if it provably underwrites some aspect of the organization's core functionality.

A rigorous set of protection requirements are specified for just those assets that directly enable the organizational mission. Rigor is defined as the ability to resist any known or conceivable method of attack. Relevant stakeholders are assigned to supervise and maintain each asset. Effective communication linkages are established between those stakeholders and documented. Then the protection requirements, access links, and the requisite permissions are enabled as a coherent set of electronic and behavioral controls. This includes methods for initiating, planning, executing, and following up/remediating active behaviors for the purposes of systematic control. It also includes the definition and assignment of all roles and responsibilities for every participant in the supply chain, customer, supplier, and integrator. It also includes the best practices for documentation and reporting of control information to appropriate sources.

Chapter 5: Control Design and Deployment

The only way to ensure proper implementation of a critical process is through design. Design deploys the controls required to ensure a critical asset. This is a strategic governance activity. Design creates an infrastructure of substantive controls to effectively satisfy its stated mission, goals, and objectives. Therefore, this phase identifies the explicit control objectives for each critical asset.

The architectural development process prioritizes those objectives and implements targeted control actions to most effectively achieve priority objectives. Then it analyzes and assesses the deployed control set to ensure that the resultant infrastructure satisfies the critical purpose. If documented control objectives are not satisfied, this process undertakes the necessary analysis to modify controls or plug gaps.

Chapter 6: Control Assessment and Assurance

Cyber resilience must be assured. This is a testing and assurance function that characterizes the explicit level of control performance against the protection goals. The intention is to be able to say with assurance that the aggregate controls for any given acquisition are effective given the aims of the organization.

Operationally, this should take place within a defined reporting and decision-making structure. This is an assessment process comparable to the systematic monitoring and adjustment processes that most organizations employ to assure the effective performance of its functions. Because the overall purpose of assurance is to produce a trustworthy assurance outcome, the outcome of the Assurance and Evolution phase is continuous assurance of process correctness.

Chapter 7: Recovering the Non-Priority Assets

Organizations need to understand how resilient its critical services are. This is essentially the continuity management principle. The goal of recovery planning is to ease the impact of disruptive events by using well-established plans to ensure predictable and consistent continuity of the critical services. To do this, the critical service's operating environment is studied to identify all potential failure modes, and then a proper strategy to recover from all possible breakdowns or disruptions is devised.

The goal is to create a complete and consistent recovery process that will address all conceivable types of system compromise. The plan for incident recovery must be explicit for every asset, and lessons learned are compiled to develop improvement strategies. This requires an operational plan capable of identifying, analyzing, responding to, escalating, and learning from all adverse incidents in addition to a well-defined process for assigning roles and responsibilities and managing and tracking resolutions.

Chapter 8: Ensuring a Continuously Cyber-Resilient Organization

The goal of recovery planning is to ease the impact of disruptive events by using well-established plans to ensure predictable and consistent continuity of the critical services. This is essentially the continuity management principle. Organizations need to maintain the assured functioning of all its services. This is an infrastructure evolution process.

To do this correctly, the critical service's operating environment is studied to identify all potential failure modes, and then a proper strategy to recover from all possible breakdowns or disruptions is devised. The goal is to create a complete and consistent infrastructure of controls that will achieve the purposes of cyber resilience.

The plan for development of this infrastructure must be explicit for every asset, and lessons learned are compiled to develop improvement strategies. This requires an operational plan capable of identifying, analyzing, responding to, escalating, and learning from all adverse incidents.

Expectations

This book presents an approach that is meant to implement a state of cyber resilience as a real-world condition. This is a business-level activity. Therefore, there are no expectations about specialized technical knowledge. All readers will learn how to design and evolve a cyber-resilient architectural process for a given organization as well as how to maintain a state of cyber resilience in the day-to-day operation of the business. After reading this, the reader will know how to ensure a stable state of systematic cyber resilience within their organization as well as evolve the protection scheme to continue to appropriately address the threat environment.

At the end of this book, the reader will be able to

1. create, sustain, and evolve a cyber-resilient organizational infrastructure;
2. define and evaluate control arrays to ensure all assets of critical value;
3. ensure full and complete recovery of noncritical assets in the timeliest and most effective way possible.

Authors

Dan Shoemaker, PhD, is principal investigator and senior research scientist at the University of Detroit Mercy (UDM) Center for Cyber Security and Intelligence Studies. Dan has served 30 years as a professor at UDM, with 25 of those years as department chair. He served as a cochair for both the Workforce Training and Education and the Software and Supply Chain Assurance Initiatives for the Department of Homeland Security, and was a subject matter expert for the National Initiative for Cybersecurity Education (NICE) Workforce Framework 2.0. Dan has coauthored seven books in the field of cybersecurity and has authored over one hundred journal publications. Dan earned his PhD from the University of Michigan.

Anne Kohnke, PhD, is an Associate Professor of Information Technology at Lawrence Technological University. After a 25-year career in IT, Anne transitioned from a Vice President of IT and Chief Information Security Officer (CISO) position into full-time academia in 2011. Anne's research is focused in the area of cybersecurity, risk management, threat modeling, and mitigating attack vectors. Anne received her PhD from Benedictine University, and has coauthored four other books in the cybersecurity discipline.

Ken Sigler is a faculty member of the Computer Information Systems (CIS) program at the Auburn Hills campus of Oakland Community College in Michigan. His primary research is in the areas of software management, software assurance, and cloud computing. He developed the college's CIS program option entitled "Information Technologies for Homeland Security." Until 2007, Ken served as the liaison for the college to the International Cybersecurity Education Coalition (ICSEC), of which he is one of three founding members. Ken is a member of IEEE, the Distributed Management Task Force (DMTF), and the Association for Information Systems (AIS).

Chapter 1

It's Time for a New Paradigm

Following this chapter, the reader will understand

1. the role and importance of cyber resilience in protecting organizations;
2. the differences between cybersecurity and cyber resilience;
3. the standard steps of the cyber-resilient approach;
4. the concerns and issues associated with our cybersecurity;
5. the general structure and intent of a cyber-resilient architecture;
6. the large steps to implement formal cyber-resilient processes.

Introduction to the Book

Two decades of data make it clear that conventional cybersecurity doesn't work (Symantec, 2014; Trend-Micro, 2015; PRC, 2016). Hence, this book offers an entirely new and different approach, one that is both resource efficient and one which ensures that the organization will continue to survive, no matter how destructive the attack. We call this approach "cyber resilience." Cyber resilience is not the same thing as cybersecurity. Cyber resilience ensures the absolute protection of just those functions that are vital to the organization's survival.

This chapter will introduce the general principles and concepts of cyber resilience as well as the standard methodology and contextual activities that guide the implementation of a strategically sound cyber-resilient architecture. We will detail the fundamental phases involved and the best practices that must be implemented in each of the phases of a classic cyber resilience process. These phases build upon each other in a collective fashion and proper execution of each is integral to the assurance of organizational resiliency.

Organizational resilience is important because the increasing presence of advanced cyber threats makes it inevitable that every organization will ultimately be targeted (OAS, 2015). Cyber resilience recognizes that there are too many cutting-edge hacking tools to prevent sophisticated attackers from finding the cracks in even the most robust cybersecurity perimeter (Lois, 2015). Thus, there is a need for a new paradigm.

Cyber resilience requires the organization to spend whatever it takes to develop a well-defined and explicit set of controls to ensure the survival of just those critical elements that cannot be subjected to compromise. The controls must assure provable protection of core functionality and the various interdependencies in the enterprise's ecosystem (EY, 2014). The concept of cyber resilience goes far beyond the classic boundaries of better access controls (EY, 2014). Instead, organizations establish a "cyber resilience strategy and architecture" that give them the ability to withstand and recover rapidly from disruptive events (EY, 2014).

Practically speaking, the best argument for cyber resilience is that it concentrates resources where they will make the most difference. This is particularly germane to national security in that any attack on an infrastructure element threatens a lot more than simple business processes (Conklin, Shoemaker, Kohnke, 2017). Thus, cyber resilience is a particularly significant aspect of ensuring survival and easing recovery of the critical systems that underwrite our way of life.

In general, it is our belief that very little substantive thinking has taken place when it comes to devising a specific and generally reliable approach to protecting the nation's critical infrastructure. This is partly because there is no practical process that explicitly dictates how to reliably protect critical infrastructure components. The ideas presented here are a start toward eventually overcoming this lack of planning.

Readers will understand the ultimate reasons why cyber resilience provides more robust assurance for the organization. They will also discover the role that conventional strategic management plays in the creation and maintenance of an effective cyber-resilient architecture. Finally, the last chapter of this book will describe the practical means for operating an organization in a cyber-resilient state.

Why Cyber Resilience Is Critically Important

Count on it—the next world war is going to start with the click of a mouse, not a wave of bombers. The attacker will send a command to the computers that control key elements of our critical infrastructure, and most of our way of life will be blasted back to the 18th century. Could this really happen? Two decades of data say that it is not only possible, but that it will indeed take place. In fact, we have struggled with this problem almost from the beginning of the internet. But the trend is always in one direction: "One massive hack after another" (Gamer, 2015). Worse, data indicates that conventional cybersecurity approaches will never be able to successfully protect us (Symantec, 2014; Trend-Micro, 2015, PRC, 2016).

Cyberspace is full of adversaries. Potential actors range from state-sponsored groups through criminal enterprises to any wacko with an internet link. So, cyber-attacks on the various elements of the U.S. critical infrastructure are a daily fact of life. For instance, the Industrial Control Systems-Computer Emergency Response Team (ICS-CERT) reports that U.S. industrial control systems were attacked at least 245 times over a 12-month period (OAS, 2015). "While China, the U.S. and Russia lead the world in cyber-attacks, virtually every government engages in such attacks, and nearly every country has its share of computer hackers" (Wagner, 2017). So, forget aircraft carriers, the ability to launch a successful cyberattack makes every nation-state into a potential superpower (Wagner, 2017).

So far, the problems have been elsewhere. Perhaps the most egregious example comes from Ukraine. In December 2015, a presumed Russian cyberattacker successfully seized control of the Prykarpattyaoblenergo Control Center (PCC) in the Ivano-Frankivsk region of Western Ukraine (Wagner, 2017). The attack marked the first time that a concerted cyberattack was successfully launched against a nation's power grid (Wagner, 2017). However, Stuxnet, in 2010, might be the first instance of "a nation enforcing policy through other means," to paraphrase Von Clausewitz (Clausewitz, Chapter 1, Section 4, 1976).

Worse, the perpetrators of the Ukrainian attack were observed conducting similar exploits against the U.S. energy sector (Brasso, 2016). Although there was never any actual disruption, many experts believe that those activities were a probe for future moves on the U.S. infrastructure (Brasso, 2016). Consequently, in the larger sense, the key question is "could a catastrophic cyberattack in the U.S. infrastructure ever occur?" The National Security Agency's former Director, Mike Rodgers, made his own evaluation of the possibility of a successful attack against critical infrastructure when he said: "it's a matter of, when, not if" (Smith, 2014).

Power grids are the most frequently mentioned targets (Wagner, 2016; Brasso, 2016; Smith, 2014). This is because the interconnectedness of power grids opens them up to "cascading failures." That is, as nearby grids take up the slack for a failed grid system, they overload and fail themselves and cause a chain reaction. Rogers says that such attacks are part of "coming trends" in which so-called zero-day vulnerabilities in U.S. cyber systems are exploited (Smith, 2014).

The reason why the protection of our national infrastructure is so critically important is that a major exploit, like a successful cyberattack on the electrical grid, could leave the U.S. cloaked in darkness, unable to communicate and without any form of 21st century transport. It would likely kill many thousands of citizens, perhaps millions, either through civil unrest, failure of public systems, or mass starvation (Brasso, 2016; Maynor and Graham, 2006).

Many experts believe that the cyberwar began in 2003 (Wagner, 2017). This was when the Northeast (U.S.) blackout occurred and caused 11 deaths and an estimated

$6 billion in economic damages (Wagner, 2017). After that attack, SCADA (supervisory control and data acquisition) attacks occurred in the UK, Italy, and Malta, among others. According to Dell's 2015 Annual Security Report, cyberattacks against infrastructure systems doubled in 2014 to more than 160,000 (Wagner, 2017).

Infrastructure Is the Target

Infrastructure systems are diverse. This diversity and the criticality of the sensors and controllers that comprise a typical infrastructure system make them tempting targets for attack. Therefore, there have been long-standing concerns about the overall digital infrastructure being vulnerable to cyberwarfare and cyberterrorism attacks (Eisenhauer et al., 2006; Nat-Geo, 2017). Nevertheless, notwithstanding the disastrous nature of cyberattacks on digital targets, none of the industries in our current national infrastructure have developed coherent plans or effective strategies to protect themselves (Brasso, 2016). This is the reason why there is increasing interest in a coherent model for defending the critical infrastructure against cyberattack (Symantec, 2014; EY, 2014).

The approach we are going to discuss here is particularly suited to ensuring the continuing survivability of infrastructure systems. This is because the focus is on maintaining core functionality rather than protecting data. The idea is to only lightly defend less critical or peripheral elements while ensuring the survival of the system. This strict emphasis on survivability versus data protection is the reason why cyber resilience, versus cybersecurity, is the approach of choice for critical systems.

A New Paradigm for Ensuring Our Way of Life

This book offers an entirely new and different paradigm. Cyber resilience ensures the absolute security and reliability of just those critical functions, which the organization needs to continue to survive and carry out its mission. Carl von Clausewitz sums up the role of strategy in this way: "Strategy is the necessary response to the inescapable reality of limited resources" (Clausewitz, 1989). In short, no General ever has the luxury of overwhelming numbers or unlimited resources. So, s/he needs to adopt an approach that is likely to succeed, given the assets that are available at the time of battle. In this respect, Clausewitz posits that a successful strategy finds the most advantageous point to concentrate all the resources necessary to achieve the primary goal, which is to win the battle even if some of the lesser objectives are not achieved (Clausewitz, 1989).

Cybersecurity is the inheritor of the old information assurance mission. Accordingly, cybersecurity is still based around creating and ensuring a protection perimeter. This perimeter is shaped by assuring all logical points of access to the protected space that lies within that boundary. Since the protection perimeter of even a small organization can involve numerous points of access, electronic, physical and human, that task normally requires an extensive resource commitment to be even be remotely successful.

Whereas, cyber resilience only ensures those organization elements that are deemed critical to system survival. The requirement to maintain the functioning of a few critical components is less resource intensive than the need to ensure the confidentiality, integrity, and availability of all assets within secure space. Therefore, cyber resilience is much more resource efficient. The narrowing of scope allows protection measures to be concentrated onto a far smaller attack surface, which notionally ensures more effective protection for the things that simply can't be allowed to fail.

Operationalizing Cyber Resilience: Saltzer and Schroeder's Principles

Cyber resilience is founded on classification, prioritization, and comprehensive strategic policy-based deployment of a rigorous set of real-world security controls (Symantec, 2014). Cyber resilience requires the creation of a set of well-defined processes, which react to penetrations of the organizational perimeter by locking down the asset they are designed to protect (US-CERT, 2016). These protection processes are both electronic and behavioral in focus and they are designed to protect key assets as well as ensure optimum recovery of the overall system in the event of successful attack (Symantec, 2014).

Saltzer and Schroeder arguably laid down the basis for cybersecurity design in their founding principles (Saltzer, 1974). The concept of cyber resilience hinges on four of Saltzer and Schroeder's lesser-known principles. Most people in cybersecurity know about and practice design concepts such as Least Privilege, Complete Mediation, Separation of Duties, and Psychological Acceptability. Figure 1.1 lists four principles that are generally not as prevalent and underlie the cyber resilience approach (Saltzer, 1974):

1. *Economy of mechanism*: Keep the design as simple and small as possible.
2. *Work factor*: The cost must be greater than the potential attacker is willing to commit.
3. *Least common mechanism*: Minimize the amount of mechanism common to all users.
4. *Compromise recording*: Reliably record the actions of a compromise.

Economy of mechanism advises the construction of simple strongpoints around critical assets rather than basing the protection on the complexities of comprehensive

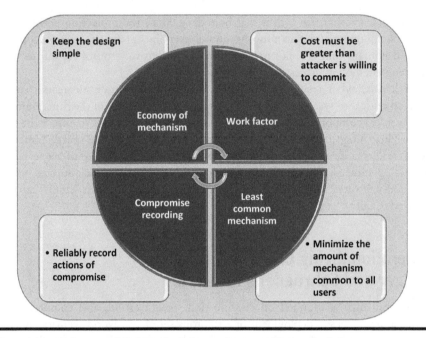

Figure 1.1 Saltzer and Schroeder's lesser-known cyber principles.

perimeter access control. In conventional military tactics, a strongpoint is a key position, which is very difficult to overrun or avoid. With respect to illustrating the difference between cyber resilience and cybersecurity, perhaps the best practical example would be that a strongpoint is like locking critical assets in a safe rather than protecting them by assuring access to the building they are in, which is perimeter access control.

In military doctrine, strongpoints are arrayed in mutually supporting, defense-in-depth mesh arrangements, called hedgehogs, rather than formed into a contiguous line of increasingly rigorous perimeter access controls. The hedgehog arrangement implements the principle of *least common mechanism* in that strongpoint defenses are tailored to just the threats affecting the protection target. This allows for very straightforward simplicity in the design. It also allows the organization to maximize security resources by building up the capabilities of only the protection for the critical features rather than diffusing the investment by attempting to protect everything.

Most importantly, if the strongpoint protecting the critical assets are robust enough, they will be too expensive for the attacker to assault. So, the attacker will be shepherded to more vulnerable targets and that brings us to the work factor principle in operation.

Work factor also serves to maximize the defender's rapid response capability. In effect, organizations will be able to rapidly concentrate their resources at the point of attack, knowing that the essential functions are protected. Since the critical system protection, e.g., strongpoint positions, will be bypassed due to their impossibly high work factor rate, the organization will be able to concentrate its resources on the recovery of noncritical functions and information.

Finally, the organization will be able to deploy and conduct the most effective recovery possible because information from prior incidents will be available to planners and responders to help optimize the response. This is the intent of the *compromise recording* principle. The effect of an attack on a nonessential resource can be minimized through lessons learned from prior attacks. Additionally, the road to recovery can be planned because the execution and outcomes of the attack have been recorded for study.

Tactics One and Two: Economy of Mechanism and Work Factor

Cyber resilience involves the formulation of well-defined strategies and the implementation of rigorous strongpoint countermeasures, which are designed to inflict unsustainable work factor requirements on an attacker. Consequently, the most important thing the organization needs to know is: "Exactly how many strongpoints will I need to build and exactly how much investment will be required to make each strongpoint infeasible to attack?" It should be possible to identify all those organizational functions that are too critical for the organization to lose and still stay in business. This involves a strategic planning process and accordingly, the organization should then concentrate sufficient resources to ensure that those specific protection targets always demand too great an investment on the attacker's part. Thus, cyber-resilient architectures are founded on making decisions about what that organization can't afford to lose and then ensuring that it doesn't lose them.

The success of a cyber-resilient defense hinges on defining protections for critical assets that are too costly for the attacker to break. These protections don't have to ensure absolute and unquestionable security, but they DO have to assure enough protection that attackers will find the resource investment unpalatable, and thus move on to another target. Since there is likely to be a much greater number of soft targets, e.g., nonessential assets versus hard ones, cyber resilience defenses will not require as great an investment as a strategy that is aimed at protecting all the resources, soft and hard, inside a perimeter.

Tactic Three: Least Common Mechanism

To implement a proper strongpoint defense, the relationship between the system's critical assets must be identified and labeled. This is necessary to establish the precise state of dependencies and interdependence of objects within the system. More importantly, the interface between the users of those assets must be well understood and characterized to implement control. Next, a broad-spectrum risk assessment ought to be performed for each of the identified system interdependencies and user accesses. The idea is to obtain full situational awareness, both in terms of critical asset interactions as well as the potential threats arising from them. Using the full situational awareness, a provably effective control response must be deployed for each of the critical assets. Resources are focused on assuring only those components that are designated as critical. This is primarily an engineering design exercise, driven by precise knowledge of the components and their interrelationships. The resources that are left over after all critical asset dependencies are ensured are then allocated to protection and recovery of the rest of the system.

Since no single function operates separately from all the other critical functions, the resilience must be baked into the architecture in such a way that critical functions cannot be accessed by a backdoor. This is a pure design/control deployment exercise. Nevertheless, resilience of the critical asset control design and deployment needs to be confirmed correct. This is a classic testing and assurance function that periodically characterizes the effectiveness of critical control performance against stated mission goals.

Tactic Four: Compromise Recording and Strategic Recovery Planning

Well-defined processes need to be established to ensure that all data obtained from both attacks and compromises is recorded for analysis and planning. The aim is to ensure that all system functions are fully restored within requisite parameters, based on a method or plan. This requires a suitable array of evidence gathering review and testing processes and metrics sufficient to evaluate any form of compromise for future planning (Bradford, 2017).

The aim is to allow the organization to wholly understand its digital environment. This allows it to build the most effective rapid response team and design well-defined scenarios for how each attack will be managed. The main reason why compromise recording is so effective is that 98% of cybersecurity incidents fall into nine basic attack patterns (Verizon, 2018) as shown in Figure 1.2:

1. Denial of Service
2. Privilege Misuse
3. Crimeware
4. Web App Attacks
5. Physical Theft and Loss
6. Miscellaneous Errors
7. Cyber-Espionage
8. Point of Sale
9. Payment Card Skimmers

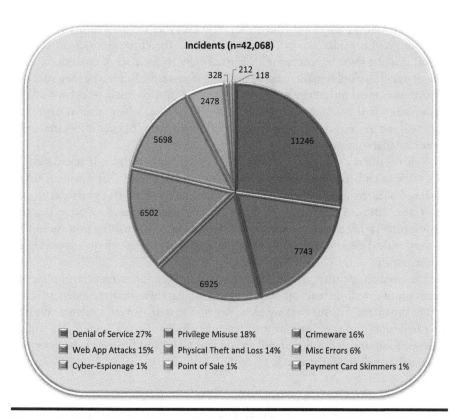

Figure 1.2 Percentage and count of incidents per classification attack patterns.

These basic attack patterns can be studied, documented, and a response can be crafted to ensure the most effective resolution for a given known situation. The ability to dynamically respond to exploits, based on lessons learned, is crucial to keeping costs down. If this is done correctly, the organization will have both security and cost efficiency.

Cyber Resilience versus Cybersecurity

The underlying assumption is that redesigning or updating the organization to a cyber-resilient architecture will make attacks on organizational functions less likely to succeed. In addition, the architecture will minimize the consequences of cyber-attacks when they do succeed, increase costs and uncertainty for the adversary, and possibly act as a deterrent against future attacks. Thus, cyber-resilient organizations can "tough-out" the types of assaults that bring conventionally protected organizations to their knees.

The increasing presence of advanced, persistent cyber threats makes it inevitable that a targeted organization will be compromised. Therefore, every organization's critical systems must be assured reliable under any realistic set of conditions. Lost data or ancillary functionality can be restored afterwards. But in the long run, the organization must ultimately continue to meet its mission goals. A cyber-resilient organization will continue to fulfill its critical purpose, where normal organizations will fail. Accordingly, investment in building cyber resilience is investment in organizational survival.

Cyber resilience recognizes that people are the weakest links. It acknowledges that workers and security personnel will make mistakes and that intruders using advanced hacking tools will find cracks in even the most robust cybersecurity system (Lois, 2015). Accordingly, cyber resilience is baked into the "bones" of the organization rather than added as an ancillary feature. The baking process applies to every critical electronic, human, or physical element of the enterprise's operating infrastructure.

The need to identify all critical assets within that infrastructure implies the execution of a well-defined process to find and guarantee that the assets that are vital to organizational survival are identified and that an explicit understanding of their behaviors and their various interdependencies has been achieved. The aim is to devise foolproof controls that will ensure that all critical functions and assets are adequately protected from attack.

Changing the Culture of Information Protection

Implementing this radical new approach demands a change in culture. The most important first step in the process is to recognize that cyber resilience is not an electronic security discipline focused strictly on detecting network intrusions or fixing malware. The simple fact is that there is no existing evidence to support the conclusion that the enterprise will ever be 100% secure if only the electronic parts of the system are protected. There are simply too many threats in the ecosystem to assume that.

Consequently, organizations must stop viewing their operational security in terms of identification, authentication, and authorization of requests for access. Instead they need to start treating the response to threat as a comprehensive, holistic, fully integrated, strategic management solution, which is rooted in threat identification, targeted control architectures, and enterprise recovery.

Effective implementation of enterprise-wide cyber-resilience architectures requires strategic vision. It also requires day-to-day engagement of the security function across the enterprise. The former requirement is rooted in the need to integrate and coordinate cyber resilience planning with the specialized actions and collaborations that take place within the classic cybersecurity function. The role of the latter focuses more on identification, authentication, authorization, and enforcement of access and response to incidents as they might occur, while the cyber-resilient mindset guarantees the ultimate survival and continued functioning of the organization in its entirety.

Ensuring Optimum Use of Resources

Carl von Clausewitz wrote the book on war. In it, he summed up the role of strategy this way: "Strategy is the necessary response to the inescapable reality of limited resources" (Clausewitz, 1989). So, a practical approach must be devised, which is likely to succeed, given the assets that are available at the time of battle. In this respect, Clausewitz posits that a successful strategy finds the most advantageous point to concentrate all the resources necessary to achieve the primary goal, which is to win the battle even if some of the lesser objectives are not achieved (Clausewitz, 1989).

Von Clausewitz might have lived through the Napoleonic wars, but he could just as well have been writing about the present state-of-affairs in cyberspace. The hosts of threats that populate the virtual ecosystem are so numerous that it is impossible to defend against every form of attack (Symantec, 2014). The unfortunate reality is that every system can eventually be compromised and no computer is safe. Furthermore, America's electronic infrastructure is so riddled with vulnerabilities and there are so many adversaries out there that "cybersecurity" is a relative term indeed (Brasso, 2016).

Adversaries in cyberspace range from state-sponsored hackers to the professional perpetrators of cybercrime to any person with a grudge and an internet link. Moreover, because of the proliferation of adversaries and the inescapable vulnerability of the U.S. critical infrastructure, cyberattacks on its various elements are a daily fact of life. For example, the nonprofit Privacy Rights Clearinghouse reports that we have lost over one BILLION records over the past decade and these are just the losses that were REPORTED. Since most companies do not like to publicize their security failures, that number could be, and probably is, much higher.

Over time, the running average of 100 million records lost per year has been subject to some variations and the source of breaches has changed in common-sense ways as the technology has evolved. Nonetheless, the number of reported incidents rose annually from 108 in 2005 to 607 in 2013 (PRC, 2015). The important point to note is that while losses have increased at a rate of 50% per year over the same period, corporate budgets for security have increased at the rate of 51% (Symantec, 2014).

At the same time, 35% of respondents surveyed report that employee records were compromised, 31% report that customer records were either compromised or unavailable, and 29% say that internal records were lost or damaged (Hulme, 2013). Those losses happened even though security budgets increased at the rate of 51% over the same period, while annual losses of more than ten million per incident are up by 75% (Symantec, 2014).

Consequently, it is reasonably safe to assume that cyberattacks are going to happen, no matter how much money is thrown at the problem. At the same time, notwithstanding the clear recognition of the disastrous nature of a successful cyber-attack, none of the industries in our current national infrastructure have developed any form of effective standard strategy to protect themselves (Brasso, 2016). This includes the Chemical Sector; Commercial Facilities: Communications; Dams; the Defense Industrial Base; Emergency Services; Energy: Financial Services; Food and Agriculture; Government Facilities; Health Care and Public Health; Information Technology; Nuclear Reactors, Materials, and Waste; Transportation; and Water and Wastewater Systems (PPD-21, 2013, p.2).

Right now, we secure systems at the logical points of access. Even defense-in-depth, which is the modern doctrine for doing that, simply involves several increasingly rigorous perimeters. Cyber resilience suggests that, instead of investing our resources in defending the system at all points of attack, we need to devise ways to ensure that the functional assets that we cannot afford to lose are unconditionally protected. In concept, this strategy would narrow the attack surface to the point where absolute assurance within available resources could be guaranteed while reducing the impact and recovery time of nonessential assets. This selective versus universal approach to security is what sums up the difference between a *cyber-resilient* strategy and one that is based on cybersecurity.

Evildoers will ultimately compromise any system that they target (Symantec, 2014). The problem is that the universe of threats is so varied that it is impossible to predict the exact form of a virtual attack across the entire range of protection targets. America's electronic infrastructure is riddled with vulnerabilities and there are a lot of adversaries out there who are looking to exploit them (Lois, 2015; Brasso, 2016). They range from state-sponsored hackers to criminals to any person with a grudge and an internet link. So, perhaps it is time to circle the wagons in a way that is designed to protect just those things that we know we can't lose.

One would think that the well-documented presence of massive, persistent threats would cause sleepless nights among the CEO population. Yet, 84% of CEOs, 82% of Chief Information Officers (CIOs), and 78% of Chief Information Security Officers (CISOs) believe that their cybersecurity programs are effective (Hatchimonji, 2013). Given the overwhelming array of evidence, it would be difficult to conclude that we sare doing a BETTER job of protecting information. So, if system compromises are a given, maybe we need to reconsider the paradigm. Perhaps we should think in terms of reducing impact and recovery time rather than unsuccessfully trying to defend the system perimeter.

Perhaps, we need to revisit the logic behind the approach to security that we adopted long before the internet became commonplace. Right now, we defend our organizations like the seventh cavalry defended Fort Apache. Our defense is arrayed along the walls to prevent any intrusion into secure space. Penetration of those walls will give the adversary access to the space it encompasses. Therefore, the logical place to prevent that access is at the precise point where insecure space becomes secure space. So, the perimeter must be defended at all costs.

The perimeter protection approach to securing a given set of logically grouped assets holds true even in so-called defense-in-depth architectures, and it is tremendously resource intensive because the access points will only be as secure as the number of soldiers you deploy to protect them. In case of large systems that implies a lot of cavalry.

In the end, assurance depends on the level and degree of resource commitment. Cyber resilience postulates that you can't put a cop on every street corner. So, instead of throwing our resources into defending the space at the perimeter, perhaps we should forget about the concept of secure space entirely and devise ways to ensure that ONLY those assets that we don't want to lose are unconditionally protected. This selective versus universal approach to security is what differentiates a *cyber-resilient* strategy from one that is based on cybersecurity, and it requires rethinking the way we go about creating protection schemes.

As a general security approach, the more bounded set of activities that fall under the generic heading of "cybersecurity" ensures against threat from unauthorized access. In this respect, cybersecurity requires the design and deployment of comprehensive defenses on the perimeter to protect against exploitation from external or internal sources. Then, if the perimeter is breached, cybersecurity prescribes an appropriate and well-defined counterresponse for any compromise detected.

Conversely, the mission of cyber resilience is to install security controls that will only ensure the continuous effective functioning of the organization's core elements, should a compromise or security breakdown occur. In many respects, the cyber resilience concept is comparable to locking only the organization's most valuable resources in a safe in a building that is lightly protected by locks at its doors and windows.

The safe method is much more secure and much less resource intensive than trying to guarantee the absolute integrity of every wall, door, and window in the entire building. It should also be evident that defending the integrity of the perimeter of the building versus ensuring the survival of just those assets that are vital to the survival of the operation are two entirely different concepts involving two different types of resource commitment.

Cybersecurity focuses on preventing attacks, while cyber resilience focuses on assuring that only critical assets, e.g., the things that the organization can't afford to lose, are reliably assured. The rest of the multitude of potential protection targets is then arranged in a way that they can be quickly and effectively restored to a desired level of functioning.

Ideally, perimeter protection, e.g., cybersecurity, and critical asset protection, e.g., cyber resilience, can be arrayed to work in tandem. Both are important and neither works as effectively as it should if the other is missing. Nevertheless, a cyber resilience strategy delivers much more assurance value for the investment of scarce resources.

Figure 1.3   Required processes of a cyber-resilient organization.

Designing for Cyber Resilience

Cyber resilience is an emerging concept. Nevertheless, the four processes that are required to create a cyber-resilient organization are well known and common elements of conventional business operations. These processes are *asset identification, threat assessment, risk evaluation,* and *strategic planning,* and are shown in Figure 1.3.

These critical processes support prioritization decisions that are often difficult but necessary to underwrite the design, deployment, and assurance of an effective set of resilience controls. Resilience processes characterize and then seek to ensure that a well-defined and everyday operational process exists to ensure the resiliency of the enterprise's currently identified critical functions.

The goal is to assure the practical ability to minimize harm to just those organizational elements that maintain their basic functioning as well as fully recover everything else within an acceptable period. A human body analogy might be the easiest way to understand the aims of cyber resilience. People can function and even return to normalcy if they lose an arm or a leg or perhaps even a lung. But they cannot continue to function if their heart or brain is lost. Thus, there are parts of the human body that must be protected at all costs and there are others that can be put at risk if the resources aren't there to ensure them.

Obviously, decisions about what to protect and what can be recovered can be contentious and painful. But the fact remains that there are things everybody in the organization can agree need protecting and those get priority. Implementing such an approach entails an incremental strategic planning and architectural development process, in the sense that the requisite capabilities are phased into existing enterprise processes in a planned and rational fashion.

The Specific Example of SCADA

Cyber resilience was initially seen as the best way of visualizing protection schemes for the automated SCADA systems that perform the myriad functions that underwrite our daily lives. Thus, a lot of the current funding and research in cyber resilience centers on its applications in SCADA and other critical infrastructure system protection. Because all the elements of a SCADA system are essentially devoted to a given purpose rather than general use, they are also one of the best ways to visualize how cyber resilience operates.

Complex control networks are central to the day-to-day operations of each of the sectors in the critical infrastructure. But there have been long-standing concerns about the overall SCADA-based infrastructure being vulnerable to cyberwarfare and cyberterrorism attacks (Eisenhauer, 2006; Nat-Geo, 2017). This is the reason why the cyber resilience approach has gotten a lot of recent interest (Symantec, 2014; EY, 2014). The cyber resilience approach maintains core functionality at all costs. This happens without consideration of defending less critical or peripheral elements. This strict emphasis on survivability is the reason why cyber resilience versus cybersecurity is the approach of choice for critical systems.

Nonetheless, the generic concept of cyber resilience could potentially apply to every enterprise system operated by every organization in the world. Cyber resilience comprises all the steps that an organization needs to take "to prepare for and adapt to changing conditions and withstand and recover rapidly from disruptions. An organization is resilient when it exhibits the ability to withstand and recover from attacks, accidents, or naturally occurring threats or incidents" (PPD-21, 2013, p. 3).

To ensure cyber resilience, the organization must be able to incorporate operative, real-time incident management and recovery processes into a comprehensive and coordinated, organization-wide strategy for critical asset protection. The aim is to ensure business survival in the light of every conceivable cyber risk. The practical aim of cyber resilience is to allow an organization to experience a cyberattack while still maintaining critical functions.

Creating a Cyber-Resilient Architecture

Cyber resilience is not just a matter of building higher and thicker walls. Cyber resilience involves following a well-defined process to create an architecture that will attempt to mitigate all malicious or hostile incidents on the perimeter, just as cybersecurity does, while providing reasonable certainty that the attacker cannot reach or harm the organization's most critical assets.

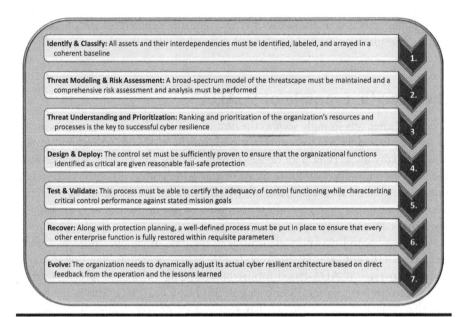

Figure 1.4   The cyber resiliency process.

The resilience control model is founded on comprehensive, risk-based strategic policy decisions. Those decisions lead to the deployment of a rigorous set of real-world controls, which are designed to control conventional access at the perimeter, protect key assets, and ensure optimum recovery time for those assets that do not have fail-safe protection. The cyber resiliency process is embodied in seven generic principles shown in Figure 1.4.

1. *Identify and Classify*: You can't protect things that you don't know exist. Therefore, all the organization's assets and their interdependencies must be identified, labeled, and arrayed in a coherent baseline of "things." That baseline describes all potential items of value and protection targets, and it is maintained under strict configuration management discipline. The activities that take place during this phase are comparable to the "Classification" activities specified in the National Institute of Standards and Technology (NIST) Risk Management Framework (NIST, 2014c). These specifications can serve as a model for an organization to design its own classification process.

2. *Threat Modeling and Risk Assessment*: Resiliency design requires appropriate situational understanding. Therefore, a broad-spectrum model of the threatscape must be maintained. All threats must be identified, understood, and a risk analysis performed to maintain up-to-date situational awareness. In that process, all known threat vectors must be identified, their interactions and impacts understood, and a value proposition developed to guide decision

making. A comprehensive risk assessment and analysis supports this activity. This must be performed for every threat scenario as it applies to the identified asset base. This process is comparable to the "Security Assessment" phase of NIST 800-53a (NIST, 2014). It guides the subsequent strategic decisions about investment. Consequently, this stage and the following stage are inextricably linked in the process of cyber resilience development.

3. *Threat Understanding and Prioritization*: This is the stage where cyber resilience and cybersecurity diverge as approaches. Because only critical assets are given the maximum assurance, ranking and prioritization of the organization's resources and processes are the keys to successful cyber resilience. Consequently, the organization should conduct a rigorous prioritization of just those assets that it absolutely can't afford to lose. These assets are selected, evaluated, and a provably effective response is deployed for each. Then the organization's resources are focused on only assuring those critical things in priority order. What is left over is then allocated to defending the rest of the organization on the same descending scale. This is comparable to the "Select" phase of the NIST Risk Management Framework (NIST, 2014). It will determine the tangible form of the protection scheme.

4. *Design/Deploy*: Resilience is "baked into" the architecture in the form of a tangibly identified and managed control set. The control set must be sufficiently proven to ensure that the organizational functions identified as critical are given reasonably fail-safe protection from an otherwise successful intrusion into controlled space. This is a design/control deployment exercise comparable to the "Implement" phase of the NIST Risk Management Framework (NIST, 2014). The outcome of this phase is a tailored set of controls that can be identified, evaluated, and tested to ensure their ongoing effectiveness, given organizational protection goals.

5. *Test and Validate*: The ongoing successful operation of the resilience control architecture needs to be assured. This assurance is sustained by a planned and scheduled oversight process. This process must be able to certify the adequacy of control functioning while characterizing critical control performance against stated mission goals. Conventional assurance methods like pen-testing apply here. This is an assessment process comparable to the "Assess" phase of the NIST Risk Management Framework (NIST, 2014). It ensures ongoing correctness of the control set and it also represents the end of the cyber resilience asset assurance process.

6. *Recover*: This stage is planned in conjunction with the control deployment activity, but it has different goals. The planning process begins when the organization reaches an agreement about the processes it will unconditionally attempt to protect. Along with the protection planning, a well-defined process must be established to ensure that every other enterprise function is fully restored within requisite parameters. This is comparable to the "Plan-Purpose-Scope-Relationship" recommendations embodied in NIST

SP 800-34—revision 1. Given that resources are frequently limited, the assets that are likely to fall under the recovery function versus protection will far exceed those that are being explicitly assured. Therefore, the scope of the Recover stage is most likely to exceed the protection operation by many magnitudes of effort.

7. *Evolve*: The one thing that is certain about the threatscape is that it is constantly evolving. New threats appear every day and old threats mutate into new forms that have not been accounted for. Therefore, the organization needs to dynamically adjust its actual cyber-resilient architecture based on direct feedback from the operation and the lessons learned from that. This is a continuous and ongoing stage that is different from the control deployment and the recovery processes. It is also much more cyclical in both planning and execution. In that respect, it is comparable to the continuing execution of the process that is outlined in the NIST Cyber Security Framework (NIST-CSF, 2014a). Because it is meant to be continuous, once this stage is attained the process of managing the evolution of the protection scheme will characterize the actual cyber resilience effort for that organization.

Presidential Policy Directive-21: The Government Weighs In

The "bend but not break" strategy of resilience is articulated in Presidential Policy Directive-21 (PPD-21, 2013). This is an important point to note because a PPD articulates the National Strategy for a given area of broad-scale policy. PPD-21 defines security as taking the steps to minimize "the risk to critical infrastructure from intrusions, attacks, or the effects of natural or manmade disasters" (PPD-21, 2013, p. 1). PPD-21 terms this new approach as "cyber resilience." Thus, the general aim of cyber resilience is to install strategies and processes that will let the U.S. Critical Infrastructure "withstand and rapidly recover from all hazards" (PPD-21, 2013).

Cyber resilience comprises all the steps that an organization needs to take "to prepare for and adapt to changing conditions and withstand and recover rapidly from disruptions. An organization is resilient when it exhibits the ability to withstand and recover from attacks, accidents, or naturally occurring threats or incidents" (PPD-21, 2013, p. 3).

From a regulatory standpoint, this directive specifically addresses cyber threats to, shown in Figure 1.5, the Chemical Sector; Commercial Facilities; Communications; Dams; the Defense Industrial Base; Emergency Services; Energy; Financial Services; Food and Agriculture; Government Facilities; Health Care and Public Health; Information Technology; Nuclear Reactors, Materials, and Waste; Transportation; and Water and Wastewater Systems (PPD-21, 2013, p. 2).

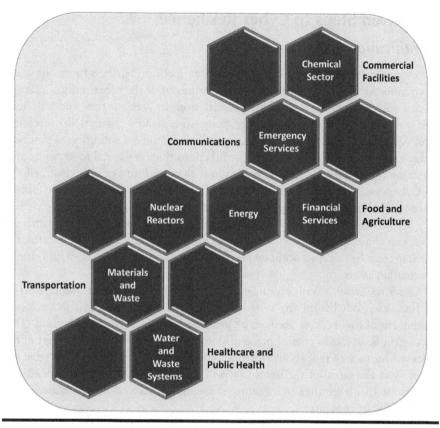

Figure 1.5 National critical infrastructure systems.

Nonetheless, the generic concept of cyber resilience could potentially apply to every enterprise system operated by every organization in the world.

To comply with this directive, the organization must incorporate operative, real-time incident management and recovery processes into a comprehensive and coordinated, organization-wide strategy. The aim is to ensure business goals in the light of every conceivable cyber risk.

The practical aim of cyber resilience is to allow an organization to experience a cyberattack while still maintaining critical functions. Consequently, a cyber-resilient organization will always be able to survive any given attack on its critical operational functions. Nevertheless, resiliency must be baked into every organizational process rather than get added on. That implies the requirement to make strategic design changes to each organization's enterprise architecture. In conjunction with the goal of protecting core functionality, the cyber resilience process also defines straightforward and practical paths to restore any lower priority functions that might have been lost in the actual compromise. The six generic stages of cyber resilience are designed to achieve those two specific goals.

The Seven Steps to Cyber Resilience

Identification and Classification

The core purpose of cyber resilience is to maintain critical business functions under all circumstances. Thus, the decisions that come out of the cyber resilience design process will determine how the business will invest its precious time and resources. This is all embodied in seven basic stages designed to create a cyber-resilient architecture. Within that architecture, the business deploys a robust collection of processes and controls to ensure that the subset of functions that is essential to its continuing operation is protected. This is the goal even if all other forms of protection fail.

Cyber assets are like no other resource. They are mostly abstract, in the form of processes and information, rather than concrete. Because cyber assets are virtual, they are invisible and cannot be precisely valued. Thus, their form and worth are subject to constant reinterpretation. So, to ensure effective assurance control, the first step must be to get an accurate picture of the composition of the organization's information assets.

Cyber resilience is built around risk assessment and control assurance of critical functions. But before any of this can be done, every item that is a relevant component of that process needs to be precisely identified and characterized. That stipulation is, in fact, a critical precondition for the performance of the rest of the process. One cannot talk about assuring something if no one understands what the target is, so the forms of the items that will be assured must be fully defined and documented. This requires an inventory of the information resource to differentiate and describe its contents.

The process of identification and labeling creates a picture of the exact form of the asset. This picture becomes the concrete point of reference for the deployment and management of the subsequent controls, and this snapshot is called a baseline. The baseline is the tangible specification of the structure of the organization's asset base. It documents the asset base and it serves as the starting point for risk analysis, prioritization, control deployment, and measurement control. The asset base comprises those things that the organization considers to be valuable. Therefore, because information is intangible the baseline becomes, in effect, the proxy for the asset itself.

The baseline identification scheme is crucial for two reasons. First, because the specific assurance countermeasures and controls will be directly referenced to the structure and content of the asset base, identification will essentially define the requirements for subsequent controls. And second, because the identification scheme constitutes the only official documentation of the asset. In the latter respect, this part of the process serves to guide the precise specification of the organization's planned and coordinated response to all relevant threats. These are assembled into a single working solution.

There are two separate steps involved in conducting identification. First, the criteria that will be used to identify the individual asset items must be agreed on and made explicit. This includes an itemization of the decision criteria that will be employed to define the various qualities of the asset. For instance, statements like: "The asset must be directly traceable to and support a business process" could be used as a basis for deciding whether a given process or item of information is valuable to the organization. Once the identification criteria are defined and publicized, it is important to have a procedure in place that will ensure that the people who are directly responsible for the actual identification and labeling process follow them in a systematic fashion.

Then each virtual asset is identified and appropriately labeled. This is essentially a documentation process and it is always associated with the business case. The labels that are used to document this representation reflect the overall architecture of the organization's asset base. This must be presented in such a way that the relationship between all component entities can be understood. That is documented and maintained by a formal status accounting function. Status accounting maintains a running account of all asset baselines and performs the routine housekeeping activities necessary to ensure that the picture is up-to-date.

In conjunction with the goal of protecting core functionality, the cyber resilience process also defines straightforward and practical paths to restore any lower priority functions that might have been lost in the actual compromise. So, in some respects, cyber resilience can be said to combine coordinated architectural development with robust, organization-wide incident response and continuity management to obtain a single effective solution.

The process begins with the classification of all the organization's virtual assets. That is a logical point to begin, since it's almost impossible to intelligently respond to a threat if you don't know what you're protecting. Right now, only 18% of organizations surveyed reported being extremely effective in their ability to target their protections (EY, 2014). The majority reported that they were minimally effective or did not know how effective they were (EY, 2014).

In many respects, identification and classification of the elements of operation are perhaps the most significant steps in creating a cyber-resilient organization. That is because the outcome of the classification process will drive every subsequent protection action. Thus, classification is the key starting point. It is an organization-wide exercise whose aim is to understand the criticality, sensitivity, and priority of all the items in the asset base.

Before any control scheme can be devised, the form of the asset must be known and categorized. The identification process entails a meticulous classification and labeling of every item placed into the security scheme. This is not a trivial requirement. It is a prerequisite for subsequent risk assessment. That is because it establishes the "day one" state of the organization's entire inventory of business functions, supporting information assets and associated technologies.

The documentation of the aggregate set of assets is termed a *baseline*. All the individual components that constitute that baseline must be explicitly identified and labeled as part of the process. A precisely defined asset baseline is an absolute prerequisite for the conduct of the rest of the process. That is because the baseline components that are analyzed, controlled, and maintained by the security controls constitute the only visible representation of the entire virtual operation.

The baseline is a concrete, formally defined structure. Classification and tagging of the asset elements are always based on their logical interrelationships. Then that baseline is maintained as a top-down hierarchy of elements that range from a view of the organization as a single entity down to a detailed understanding of the explicit items that constitute every function and resource within that organization.

The decisions that establish the baseline are made using the input of many participants. These range from the technical staff, all the way up to executive owners of a given asset. The decision making is likely to be political, and at times bloody, so the process should be designed and conducted in a way that the outcome is a rational description of priorities not influenced by external considerations.

Each item placed in the hierarchy is given a unique and appropriate label, explicitly associated with the overall structure of the information asset base itself. The label should designate and relate the position of any item to the overall family tree of items in the asset baseline. Once established, the formal baseline is kept as an organizational asset, which is fully accounted for and maintained throughout the life cycle of the security scheme.

Because information and technology continuously evolve, procedures should be defined to systematically identify and manage change. For example, something as simple as adding a new data element, for instance social security number, to a record will likely change the security status of that record. That will require the organization to update the baselines that contain that data element to reflect its new status. If the baseline is not updated in a systematic and disciplined fashion, knowledge of the form of the asset will move out of the organization's grasp, leaving it to secure things that do not exist and not securing things that do.

Thus, the baseline must maintain a continuously accurate accounting of all items. Correctness is particularly important with information because the baseline represents the actual asset itself. The problem is that baselines are dynamic, because the business itself is constantly changing. So, over time, as the form of the asset and nature of the threats change, the baseline must evolve.

This evolution is done using a disciplined process. That is because the organization's cyber assets are mainly virtual and therefore difficult to visualize. Consequently, without a distinct procedure to ensure that the evolution of the baseline is properly managed, it is likely that the necessary understanding of the asset base, maintaining the actual cyber resilience control process, will be lost.

Ensuring against that requires a plan. The goal is to establish a persistent organizational scheme to maintain accurate understanding of the shape of the asset base. This is a critical stage because it makes the virtual asset base visible to the organization at large. This scheme must integrate the organization's virtual and cyber assets into a coherent representation of everything considered worth managing and protecting. Because of that practical requirement, this part of the process is almost always guided by the business case.

The primary benefit of well-defined and up-to-date asset baselines is that they fully underwrite the recovery stage in the process, which is step six in this model. Recovery is an essential feature of cyber resilience because it specifies the organization's planned response to every potential threat scenario.

Knowledge of the baseline status prior to an incident is a critical assurance element, because it allows the organization to recover assets after a security event has taken place. The contribution of this initial phase of the process to disaster recovery is up-to-date assurance of knowledge of the information resource and its associated controls. That means it will be possible to restore normal functions, in a timely fashion, to a specified recovery point if an incident occurs.

Threat Identification and Risk Assessment

Risk assessment provides timely and accurate understanding of the threat status of all components within the asset base. That is essentially a situational awareness requirement. That is, the Risk stage employs situational awareness practices to drive the decisions about the best way to ensure critical assets, and services will continue to function as desired. The aim is to fully understand every conceivable threat in the organizational ecosystem.

The term "threat" denotes a danger or any incident, natural or man-made, that warrants explicit actions to protect a given asset against harm, to minimize disruptions of the mission. That includes natural disasters, cyber incidents, acts of terrorism, sabotage, and destructive criminal activities targeting critical components of the enterprise infrastructure (PDD-21, 2014).

The outcome of this phase is a detailed description of all known threats and their associated impacts and likelihood, sufficient to support decision making with respect to organizational priorities. No decisions about priorities should be made until the entire threat picture is fully characterized. There is the potential that some form of meaningful threat might not be missed in the initial identification process or that a new threat will appear after the original risk assessment is completed. Therefore, there also should be a comprehensive plan to guide how subsequent risk information will be included in the definitive baseline classification scheme, as it applies to any potential attack surface.

There are many threats in the organization's ecosystem. These represent a hazard to some aspects of organizational functioning. Logically then, the mitigation of each threat requires sufficient understanding to devise an effective approach to managing it. This is a complex process with lots of inherent detail required. Therefore, to conduct this phase of the process, it is necessary to identify and sufficiently classify the precise nature and operation of the threats that exist in the organization's current operating environment.

Accordingly, the organization should devise and adhere to a plan to identify, classify, and prioritize threats as they appear. All plans for any form of risk evaluation should be based on accurate and consistent standard assessment techniques. Accurate and consistent assessment is critical because management will use the assessment data to make decisions about the baseline functions that are affected and their degree of risk exposure as well as the types of controls that will have to be deployed. Those decisions subsequently drive the prioritization process. Accordingly, all the metrics involved in the risk evaluation process must be unambiguously and rigorously defined and agreed on. Those definitions can then be used to ensure that the data from the assessment process is accepted by all stakeholders.

Accuracy and consistency are critical factors because stakeholders should share a common understanding of the precise nature of the threatscape to define priorities. As a result, it is important to make certain that there is consistent knowledge of what a given assessment result means.

If the various people who are involved in the prioritization process interpret the results of the threat assessment differently, there is a potential for uncoordinated and ineffective control responses or even a failure to identify critical functions. Also, there is the issue of credibility when it comes to the data itself. If there is no clear definition of the basis for measurement then it will be hard for decision makers to rely on the data.

The activities that are involved in the identification of threat and the assessment of the risk they represent are planned and implemented in the same way as other types of organizational assessment activities. That is, the threat assessment process employs conventional risk evaluation techniques to decide about the nature of emerging threats.

These threats are normally derived and then understood through the creation of a threat model or some other form of strategic threat identification process. Even so, the goal of the operational threat assessment is to say with certainty that the organizational assets and their associated risks have been satisfactorily identified.

The operational threat assessment should provide all the information necessary to allow decision makers to identify those virtual assets that need to be managed to achieve the desired state of resilience. The aim is to provide explicit advice to the people who are responsible for making the prioritization decisions that follow in the next stage.

Threat Understanding and Prioritization

The resources that can be dedicated to asset assurance are always constrained. According to Presidential Policy Directive-21 (PDD-21), "Assets" comprise all the people, processes, technologies, and facilities required to achieve the organizational purpose. However, some assets are more critical than others. So, the criticality of every virtual asset in the organization must be ranked to decide where to best invest in protection. This is an organization-wide process involving all stakeholders. The ranking process identifies, documents, and assures only those assets that are ranked as "critical" to the organization's mission, vision, values, and purposes (PDD-21, 2014).

The exact process that will be used to assign final priority must be documented and agreed on. Unfortunately, the ranking of assets is a primarily political process. Ranking should be based on some, or all, of these criteria as shown in Figure 1.6:

1. Centrality/criticality of the function to business goals/purposes
2. Viability/likelihood/impact of known threats
3. Cost/benefit of mitigation
4. Interdependencies with other elements of the operation
5. Future assumptions about business directions
6. Competitive business intelligence

Because the decisions will involve political concerns, objective evidence is an important mechanism for supporting the decision making. The reason why objective evidence is so important is that most of the organization's system assets are abstract. Therefore, a common, well-defined objective and explicit data gathering procedure are required to ensure fundamental uniformity of comparison. This is necessary to facilitate stakeholder understanding and agreement when the hard decisions need to be made. Those methods include such activities as:

1. Threat models
2. Risk analyses
3. Inspections and audits
4. Static and dynamic testing
5. Competitive intelligence

Nevertheless, in the end, the evidence set will contain both subjectively derived and interpreted evidence, as well as the easier to accept objective findings. Subjective evidence is mainly gathered from discussions with individuals or groups of individuals within an organization. These are the stakeholders and the people who are directly responsible for the actual day-to-day performance of the system. Thus, their insights are valuable in understanding the fundamental operation of the system.

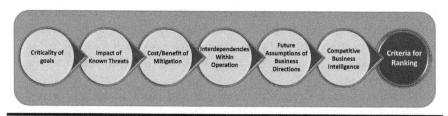

Figure 1.6 Criteria for ranking process.

Subjective evidence gathering is supported by such conventional tools as inspections, surveys, focus groups, and observations. However, all subjective evidence must be labeled as such.

Objective evidence is gathered by empirical means such as from testing and analysis. The testing can be both static and dynamic. Each method involves executing the functions of one or more assessment objects under a given set of testing conditions. The aim is to confirm that actual behavior conforms with expected behavior. Both forms of assessments are used in building the evidence necessary to support the ranking process.

The actual assignment of priority needs to be objective and involves all stakeholders, not just top-level decision makers. That is necessary to get an adequate range of perspective for prioritization decisions. Naturally, every decision maker is unique in that each has personal biases, political issues, and personal requirements. That will affect the importance assigned to a given function. Thus, the ranking process should specify the precise type and amount of evidence required to underwrite a certain level of priority.

A rigorous set of protection requirements are specified for just those assets that directly enable the organizational mission. Rigor is defined as the ability to resist any known or conceivable method of attack (PDD-21, 2014). Relevant stakeholders are assigned to supervise and maintain each asset. Effective communication linkages are established between those stakeholders and documented. Then the protection requirements, access links, and the requisite permissions are enabled as a coherent set of electronic and behavioral controls.

Design/Deploy

Design creates and deploys the controls required to ensure a critical asset. This is a strategic governance process at the top. Design creates an infrastructure of substantive controls to effectively satisfy the protection system's stated mission, goals, and objectives. Thus, this phase identifies the explicit control objectives for each critical asset. It prioritizes those objectives and implements targeted control actions to most effectively achieve priority objectives. Then it analyzes and assesses the deployed control set to ensure that the resultant infrastructure satisfies the critical purpose. If documented control objectives are not satisfied, then the Design process undertakes the necessary analysis to modify controls or plug gaps.

Design thinks through how each of the priority assets identified in the prioritization stage will be protected. The outcome of this stage is an explicit and consistent architecture of controls. The goal is to ensure that every one of the identified assets is ensured by a logical set of controls that will ensure the continuous operation of the target function.

The design dictates a practical, real-world, and provably resilient control set for each individual critical asset and then confirms that a satisfactory solution for the entire set of resilience requirements has been attained. Besides ensuring the ongoing resilience of the target functions, this step also does the necessary assurance of the external and internal interfaces of the target asset. At the end of the process, the architecture that design creates is fully and completely documented.

Qualitatively, the design process should satisfy the criteria of traceability, external and internal consistency, appropriateness of the methodology and standards employed, and feasibility. The outcome of the design process is an architectural design. This architecture must encompass all the critical elements that have been targeted for protection and address all identified threats to the ongoing functioning of the system, and all architectural components must be validated as consistent with ensuring proper resilience. This reconciliation of protection requirements must also demonstrate complete traceability to the priorities assigned to the target element and the system.

Once the control functioning has been validated as being correct and consistent with the stated assurance requirements, it is incorporated into the baseline along with the attendant element that it is assigned to protect. This baseline is then communicated to all stakeholder constituencies.

The first step in the design process establishes a top-level architecture of controls. It begins with the identification of all relevant critical assets. It incorporates each critical asset into well-defined protection modules. The internal and external interfaces of each module are also identified and documented. In addition, any interdependencies between modules are also documented and incorporated in the design. Then provably effective controls are designed and allocated. Where control integration is necessary based on dependencies, that task is also handled here. The control design is then evaluated to determine whether the outcome is consistent with all the formal protection requirements for that specific key asset. The design is also assessed to determine whether the resulting protection will ensure suitable resilience.

Finally, the designer specifies the technical requirements for each of the components itemized and ensures that those technical actions directly ensure the resilient functioning of a given protection target. The requirement for the assurance of the external and internal interfaces is the same as it is for the actual module protection design. There must be an explicit design for the control of all external and internal interface interactions including those between users as well as other components.

At this point, the specific testing and qualification requirements for each controlled key component are developed and documented. A reasonable schedule for testing each component is also defined. Once the correctness of the technical design is confirmed, it is turned over to the people who will implement it in the day-to-day operation. It is probable that there will be an existing cybersecurity system already in place. So, the resilience architecture will have to be integrated rather than installed whole-cloth. Thus, the aim of this final phase is to integrate all relevant controls into a complete system.

In that respect, then there must be a strategy for deploying the resilience controls according to the priorities expressed by the stakeholders. The plan aggregates all the relevant test requirements, procedures, data, responsibilities, and schedules for each of the targeted protection modules. The plan ensures that each module satisfies stakeholder priorities and that each of the component entities is properly integrated with all the other actions in the protection scheme. This is reported to the stakeholders for approval.

Because the integration brings together the fundamental components of an existing cybersecurity scheme, in conjunction with a new cyber resilience approach, the protection scheme must be tested against stakeholder assurance requirements. Thus, the design process also develops and documents a set of tests and test procedures for assuring the overall goals of resilience and survivability for the system as a whole. This must be formally accepted by stakeholders in the same fashion as product acceptance takes place.

Recover

Organizations need to understand how resilient the rest of its architecture is. This is essentially the recovery planning principle. The goal of recovery planning is to ease the impact of disruptive events by using well-established continuity planning principles to ensure predictable and consistent continuation of the key services as well as ensure rapid recovery of those items that are not part of the critical asset protection scheme (PDD-21, 2014).

In this respect, the term "recovery" describes the implementation of a mechanism to ensure the continuous off-site transmittal and storage of information. The term recovery also describes the processes for assigning and notifying key personnel in the event of a need to reestablish the operation. Recovery is a planned response that goes into effect if other safeguards fail. They can be both proactive and reactive in orientation. However, the single goal is to ensure the rapid restoration of full operation if an interruption occurs. Recovery is oriented toward preserving business value. Recovery processes involve all the necessary strategies and steps to ensure that the entire corporate infrastructure will survive in the event of disaster.

To do this, the critical service's operating environment is studied to identify all potential failure modes, and then a proper strategy to recover from all possible breakdowns or disruptions is devised. The goal is to create a complete and consistent process that will address all conceivable types of system compromise. The plan for recovery must be explicit for every asset, and lessons learned are compiled to develop improvement strategies. That requires an operational plan capable of identifying, analyzing, responding to, escalating, and learning from all adverse incidents and a well-defined process for assigning roles and responsibilities and managing and tracking resolutions.

The recovery process entails all the steps to ensure continuous operation. The actual methods for recovery frequently center on off-site storage and backup facilities. That is because the worst possible contingency is the loss of the primary site. Therefore, recovery plans seek to ensure the shortest possible unavailability of the physical components of the system, such as networks, servers, and storage media, as well as the continuing availability and integrity of the information that they contain.

As such, the recovery plan prescribes a precise set of actions that will have to be taken should a disaster occur. This is documented in a preparedness plan. The plan itemizes the organization's specific approach to the prevention or minimization of damage as well as all the steps it will take to secure or recover information after a harmful event.

That plan should capture a detailed and explicit process that will reestablish full functioning of the entire business operation as quickly and efficiently as possible. To accomplish this, contingency plans itemize the steps to be followed in case of a successful attack. In that respect then, the plan is based around the prioritization scheme.

Because of the critical value of the business's key assets, this plan is developed using a formal strategic planning process. The plan must characterize the operating measures that will be established to prevent avoidable disasters as well as the contingency measures that will be adopted, should one occur. It also itemizes the replacement or restoration procedures that will be utilized to ensure that the continued integrity of the organization's information is maintained.

Fundamentally, the recovery plan lays out the steps that the organization will take to ensure that its vital functions survive while recovering the rest of the assets in the most efficient and effective way possible. The information that an effective recovery plan is based on, is developed through the same threat identification and risk evaluation process that was used to implement the organization's prioritization scheme.

The requirement to develop an individual, specifically tailored recovery plan for critical assets should be kept in mind when the resilience process is first established. A separate planning effort for recovery is required because the activities to ensure absolute assurance of critical assets in the event of disaster are different from those that are required to ensure restoration of lower priority functionality.

This difference can get lost in the larger aim of securing the critical asset base. Thus, the organization should also have a plan in place to conduct a specific contingency-based recovery process as part of the resilience planning effort. The best way to ensure recovery is to build real-time survivability into the overall asset base. Whether this is part of the resilience scheme or the recovery plan, the necessity for both resilience and "recoverability" implies the integration of protection strategies with a range of proactive technologies.

The result should be a dynamic assurance solution that embodies technical elements such as firewalls and intrusion detection systems with human-based activities and services, such as real-time monitoring and management of the security technology. That degree of rigor is essential because the survival of an organization is inextricably linked to the continuing effectiveness of all its functionality, not just the critical elements.

Evolve

The Evolve stage serves as the formal basis for identifying and deploying process and technology responses and improvements across the organization. This evolution is required to meet the organization's cyber resilience goals as the threat picture changes. In this stage, measurable improvements that could increase the resilience of critical assets are identified, analyzed, and systematically deployed. The effects of currently deployed process and technology improvements are measured and the effectiveness of the selected process improvement is characterized. The five functions that must be executed mirror those of the NIST Cyber Security Framework (NIST-CSF, 2014a).

Evolution is driven by the collection and analysis of data from lessons learned about the operation of the day-to-day execution of the resilience process. Improvement recommendations are supported by data obtained from the deployment of prior process and technology controls.

Nevertheless, because this is essentially a "maintenance" activity, this type of analysis involves ongoing testing and risk estimation. Lessons learned typically involve objectively evaluating the performance of deployed processes against plans, objectives, standards, and procedures as well as the outcomes of organizational innovation and deployment processes.

The Important Role of Strategic Planning

Practically, the development and maintenance of a fully cyber-resilient organization involves strategic planning. In case of cyber resilience, the aim of strategic planning is to ensure that all the people, processes, and technologies that comprise the organization's substantive resilience response work together to achieve the purpose of cyber resilience. That requires the business to undertake a formal, organization-wide process to define, integrate, and maintain proper alignment between its business purposes and the assurance of critical functionality.

The problem with implementing a practical strategic management scheme for cyber resilience is that a lot of the key cyber resilience operations such as the information and communication technology areas and the personnel security and physical security functions operate more or less divorced from each other. In many cases, these are even separately managed units, which in some instances might even reside in disconnected parts of the organization. The role of this strategic planning function is to ensure that each of the various critical functions and functional areas work together in the most effective manner.

Thus, the organization must create and then coordinate a top-level planning process to deliberately integrate and then subsequently oversee all the security functions required to achieve the requisite level of resilience. Thus, the overall cyber resilience process embodies a defined set of inherent planning activities, which result in the tailoring of a uniform set of best practices to assure resilience goals.

The tailoring is accomplished by identifying the unique issues, problems, and criteria associated with each resilience control. The aim is to make certain that the practical technologies and controls that are installed for each critical asset achieve their intended purposes. The outcome of the tailoring process is an explicit set of controls and recovery practices that constitute the organization's current approach to cyber resilience.

In simple terms, this is a corporate strategic management responsibility. Priorities and the critical asset protection strategy dictate the requisite actions for the organization's cyber resilience function as well as how those activities will be carried out. The strategic management plan specifies the details of the control set that will comprise the cyber resilience function as well as their associated resource requirements.

In this respect, the plan provides a definition of the assurance practices for each critical component as well as a statement of the resource commitments that will be required to assure the successful operation of that function. In addition to resources, the plan also specifies the administrative model that will be used to assure proper management oversight.

The outcome of the resilience planning process is an integrated set of organization-wide security controls. It probably goes without saying that, to be effective, this complex and diverse set of actions must be coordinated. That coordination normally requires a company-wide process. The process itself must be capable of ensuring the continuous protection of critical functions and assets.

As with all formal control functions, the resilience process is composed of a distinctive set of rationally derived and logically interacting control behaviors, which are deliberately designed to achieve assured protection. These components are called "controls" because they enforce specific outcomes. Controls, in their real-world incarnations, are specifically designed policies, procedures, and/or work practices. In their practical forms, controls create a set of coherent "behaviors" that the organization then carries out in a systematic fashion.

The activities that are inherent in a control system interact to achieve the desired state of resilience. As such, the aim is to provide a detailed specification of the actions that are needed to achieve a particular given outcome. In conjunction with the chief aim of regulating behavior, the outcomes of those controls should be specified at a level sufficient to support the specific measurement purposes of a cyber-resilient organization. This all has to be planned. That is a formal activity, which is conducted by the organization as part of its overall, corporate strategic planning process. The plans that result from this process dictate all the standard actions of the organization's cyber resilience function as well as how those actions will be performed.

The strategic management plan specifies the general control objectives for the cyber resilience function as well as the associated resource requirements. In that respect, the plan specifies each critical asset in the cyber resilience scheme as well as an officially sanctioned statement of the resource commitments that will be made to that function. In addition to resources, the plan also specifies the oversight model that will be used to ensure proper operation of the controls.

Because resilience is a continuous state, not a project, the cyber resilience strategic plan must be successively refined over time as situations change. The plan has to provide explicit direction. So, it must define security control actions in concrete terms.

In order to tease out and specify a valid set of explicit control behaviors, the plan is developed and documented using a top-down process. That process starts with the critical priorities at the level of general concept and it is refined down to explicit control actions.

In essence, each general security requirement is broken down into specific procedures. From there, the procedures are given explicit behavioral descriptions. The aim is to provide a detailed description of the specific actions to be taken in the case of attack. That detailed description is necessary in order to assure the continuing appropriateness of the control actions.

Nevertheless, the plan should also communicate an explicit understanding of how the protection of the various critical assets will be coordinated within the general cyber resilience concept. That specification includes all the relevant relationships between components.

Once the general framework of controls is established, each of the inherent control actions that is required to ensure a satisfaction of the resilience aims is documented, assessed, and improved over time. That definition and documentation process is labor-intensive. But the effort will ensure that the resilience process, as a whole, is both overseen and effectively managed.

It probably goes without saying that in order to be effective, this complex and diverse set of behaviors and technologies has to be coordinated. That coordination normally requires some sort of company-wide process. The process itself has to be capable of ensuring a continuously appropriate relationship between the operational elements of the resilience function and the general vision, mission, philosophy, and cultural values of the organization. Thus, the strategic security management function ensures that the integrated set of organization-wide security behaviors are correctly performed in conjunction with the execution of the associated technological countermeasures.

The best practical justification for an organization-wide resilience process is that it ensures the optimum use of resources. That is, if all of the organization's cyber resilience controls are provably effective in the protection of the right set of targeted critical assets then none of the resources that are allocated to cyber resilience will be wasted. For that reason alone, the ability to establish and maintain good strategic alignment between the cyber resilience control function and corporate goals represents good investment.

Creating Practical Cyber-Resilient Controls

In order for cyber resilience to work properly it has to be a company-wide initiative, and the only people with the authority to enforce broad-scale strategic decisions like that are the policy makers of the organization. Therefore, the responsibility for ensuring overall implementation and operation of the cyber resilience function has to be vested with the company's top-level leaders.

Additionally, because the CIO and the CISO represent the leadership of the technology function, one or both of them has to be accountable for evolving the formal linkages between the resilience function and the conventional information and communication technology operation. The goal is for the technical leadership as well as the conventional general business leaders to participate together in the development of effective protection of the organization's critical assets.

Effective coordination of the cyber resilience control function starts with the creation of an enterprise-wide process whose sole purpose is to integrate the domains of business and information technology resilience into a single unified protection concept. A critical element is ensuring that integration is the ability to ensure effective coordination between all the people who are stakeholders in the process.

All organizations rest on critical processes. It is a given that the organization will fail if any of those processes are harmed. Cyber resilience is ensured by the implementation of a well-defined set of control objectives for each critical asset. This requires an appropriately detailed specification of all the behaviors needed to address each threat.

This specification documents the rationale for why a particular set of behaviors was chosen. The rationale details the behaviors and associated outcomes of the operation of the control. In addition, the rationale explicitly states the measures that will be employed to determine whether those behaviors met or did not meet expectations regarding the assurance objective. Once that entire set of behaviors has been tested, they are ready to be installed to protect the critical function that they have been designed to ensure.

The behaviors must be well defined and documented. That is because, by specifying an explicit set of explicit behaviors, it is possible for overseers to evaluate whether the actions specified in those behaviors are being properly executed. For that purpose, objective measures should be defined for each control objective. The outcomes of those measures will also allow managers to evaluate the associated costs. Since there are cost implications in implementing any control, the cost benefit of any required behavior should be evaluated. Because of that requirement, each control objective must be stated in precise measurement driven terms. Those measures should be able to tell management whether a specific control has achieved its assurance objectives.

Moreover, because there are never enough resources available to implement all the potential controls, it is important that the effectiveness of the control in achieving stated priorities is measured on a regular basis. Obviously, organizations can set their own criteria. However, a standard set of seven universally desirable characteristics, shown in Figure 1.7, are typically used to evaluate the general effectiveness of a control.

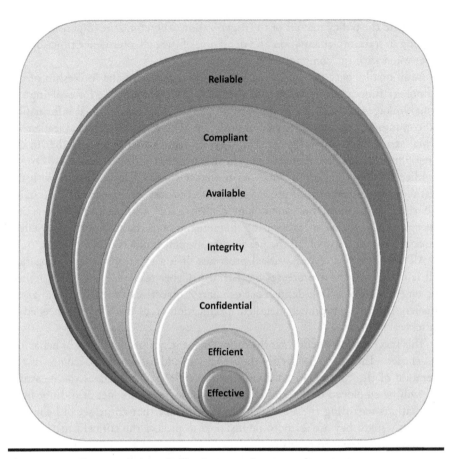

Figure 1.7 Characteristics of effective security controls.

1. *Effectiveness*—The assurance behavior must be shown to be pertinent to the stated resiliency purpose, as well as operate in a timely, correct, consistent, and usable manner.
2. *Efficiency*—Each control must operate in the most productive and economical way possible, which is sometimes called "security of mechanism."
3. *Confidentiality*—The security control must ensure that sensitive information is protected from unauthorized disclosure or access.
4. *Integrity*—The accuracy and correctness of information must be assured in accordance with the values and expectations of the business purpose.
5. *Availability*—The control ensures that information is available when needed by the business process.
6. *Compliant*—The control assures that all information and information processing comply with laws, regulations, or contractual arrangements associated with the business process.

7. *Reliable*—The control is capable of providing protection consistent with resilience goals.

Because of the potential number and scope of the controls in the control set, the actual implementation of their behaviors is hierarchical. It starts with the formulation of desired outcomes at the top and it ranges down to the specification of very precise behaviors at the day-to-day operational end of the spectrum. In an approach of this type, the organization follows an explicit series of stages, which are designed to address the threats to a given critical asset, which have been identified through whatever threat assessment that might have been performed. The practical outcome of this implementation process is a tangible cyber resilience system.

The actual steps that are taken to implement such a system typically fall into three phases. The first phase involves the establishment of a relevant set of security goals for the resilience function as a whole. In the second phase, the organization establishes the requisite level of assurance for the control set. Factors that might enter into decisions about the level of assurance include the level of criticality and the commensurate degree of threat.

Accurate boundary setting is a particularly important aspect of this part of the process, since there is an obvious direct relationship between the resources that will be required to establish the desired level of security and the number of critical assets that must be assured. Then a specific set of controls is implemented. These controls are expressed as explicit, well-defined behaviors. Those behaviors are referenced to the stated priority of the function or asset that they protect and they are normally captured in a standard documentation set.

The implementation of these controls is normally done in priority order. That process is iterative in that over time, the organization evaluates and fine-tunes the effectiveness of the control set. Alterations to the control set are based on evaluations of the performance of each individual control over time.

Chapter Summary

This chapter introduced the general principles and concepts of cyber resilience as well as the standard methodology and contextual activities that guide the implementation of a strategically sound cyber-resilient architecture. We will detail the fundamental phases involved and the best practices that must be implemented in each of the phases of a classic cyber resilience process. These phases build on each other in a collective fashion and proper execution of each is integral to the assurance of organizational resiliency.

The goal was to help the reader understand the difference between the concept of cyber security and cyber resilience. It will point out the generic threat factors that should be considered, as well as describe how each of these factors are addressed by each of those concepts. The reader will discover how the application of a formal process for cyber-resilient architecture development can establish a much more effective cyber defense at much less of an investment in scarce organizational resources.

Cyber resilience recognizes that people are the weakest links. It acknowledges that workers and security personnel will make mistakes and that intruders using advanced hacking tools will find cracks in even the most robust cybersecurity system. Accordingly, cyber resilience is baked into the "bones" of the organization rather than added as an ancillary feature. The baking process applies to every critical electronic, human, or physical element of the enterprise's operating infrastructure.

The need to identify all critical assets within that infrastructure implies the execution of a well-defined process to find and guarantee that the assets that are vital to organizational survival are identified and that an explicit understanding of their behavior and their various interdependencies has been achieved. The aim is to devise foolproof controls that will ensure that all critical functions and assets are adequately protected from attack.

Right now, we secure systems at the logical points of access. Cyber resilience suggests that instead of investing our resources into defending the system at all points of attack, we need to devise ways to ensure that the functional assets that we cannot afford to lose are unconditionally protected. In concept, this strategy would narrow the attack surface to the point where absolute assurance within available resources could be guaranteed while reducing the impact and recovery time of nonessential assets. This selective versus universal approach to security is what sums up the difference between a *cyber-resilient* strategy and one that is based on cybersecurity.

In the end, assurance depends on the level and degree of resource commitment. Cyber resilience postulates that you can't put a cop on every street corner. So, instead of throwing our resources into defending the space at the perimeter, perhaps we should forget about the concept of secure space entirely and devise ways to ensure that ONLY those assets that we don't want to lose are unconditionally protected. This selective versus universal approach to security is what differentiates a *cyber-resilient* strategy from one that is based on cybersecurity and it requires rethinking the way we go about creating protection schemes.

Cyber resilience is an emerging concept. Nevertheless, the three processes that are required to create a cyber-resilient organization are well known and common elements of conventional business operation. These processes are asset identification and threat analysis, value prioritization, and strategic planning.

These approaches support prioritization decisions that are often difficult but necessary to underwrite the design, deployment, and assurance of an effective set of resilience controls. Resilience processes characterize and then seek to ensure that a well-defined and everyday operational process exists to ensure the resiliency of the enterprise's presently identified critical functions.

The goal is to assure the practical ability to minimize harm to just those organizational elements that maintain its basic functioning as well as fully recover everything else within an acceptable period. Obviously, decisions about what to protect and what can be recovered can be contentious and painful. But the fact remains that there are things everybody in the organization can agree need protecting and those get priority. The aim is to ensure business survival in the light of every conceivable cyber risk. The practical aim of cyber resilience is to allow an organization to experience a cyberattack while still maintaining critical functions.

Consequently, a cyber-resilient organization will always be able to survive any given attack on its fundamental business functions. Nevertheless, resiliency must be baked into every organizational process rather than get added on. That implies the requirement to make strategic design changes to each organization's enterprise architecture, which results in a practical array of newly developed and integrated controls. That resilience control model is founded on comprehensive, risk-based strategic policy decisions. Those decisions lead to the deployment of a rigorous set of real-world controls, which are designed to control conventional access at the perimeter, protect key assets, and ensure optimum recovery time for those assets that do not have fail-safe protection. The cyber resiliency process is embodied in seven generic principles. Those are:

1. *Identify and Classify*: You can't protect things that you don't know exist. Therefore, all the organization's assets and their interdependencies must be identified, labeled, and arrayed in a coherent baseline of "things." That baseline describes all potential items of value and protection targets, and it is maintained under strict configuration management discipline. The activities that take place during this phase are comparable to the "Classification" activities specified in the NIST Risk Management Framework (NIST, 2014). These specifications can serve as a model for an organization to design its own classification process.

2. *Threat Modeling and Risk Assessment*: Resiliency design requires appropriate situational understanding. Therefore, a broad-spectrum model of the threat-scape must be maintained. All threats must be identified, understood, and a risk analysis performed to maintain up-to-date situational awareness. In that process, all known threat vectors must be identified, their interactions and impacts understood, and a value proposition developed to guide decision making. A comprehensive risk assessment and analysis supports this activity. This must be performed for every threat scenario as it applies to the identified asset base. This process is comparable to the "Security Assessment" phase of NIST 800-53a (NIST, 2014). It guides the subsequent strategic decisions about investment. Consequently, this stage and the following stage are inextricably linked in the process of cyber resilience development.

3. *Threat Understanding and Prioritization*: This is the stage where cyber resilience and cyber security diverge as approaches. Because only critical assets are given the maximum assurance, ranking and prioritization of the organization's resources and processes are the keys to successful cyber resilience.

Consequently, the organization should conduct a rigorous prioritization of just those assets that it absolutely can't afford to lose. These assets are selected, evaluated, and a provably effective response is deployed for each. Then the organization's resources are focused on only assuring those critical things in priority order. What is left over is then allocated to defending the rest of the organization on the same descending scale. This is comparable to the "Select" phase of the NIST Risk Management Framework (NIST, 2014). It will determine the tangible form of the protection scheme.

4. *Design/Deploy*: Resilience is "baked into" the architecture in the form of a tangibly identified and managed control set. The control set must be sufficiently proven to ensure that the organizational functions identified as critical are given reasonably fail-safe protection from an otherwise successful intrusion into controlled space. This is a design/control deployment exercise comparable to the "Implement" phase of the NIST Risk Management Framework (NIST, 2014). The outcome of this phase is a tailored set of controls that can be identified, evaluated, and tested to ensure their ongoing effectiveness, given organizational protection goals.

5. *Test and Validate*: The ongoing successful operation of the resilience control architecture needs to be assured. This assurance is sustained by a planned and scheduled oversight process. This process must be able to certify the adequacy of control functioning while characterizing critical control performance against stated mission goals. Conventional assurance methods like pen-testing apply here. This is an assessment process comparable to the "Assess" phase of the NIST Risk Management Framework (NIST, 2014). It ensures ongoing correctness of the control set and it also represents the end of the cyber resilience asset assurance process.

6. *Recover*: This stage is planned in conjunction with the control deployment activity, but it has different goals. The planning process begins when the organization reaches an agreement about the processes it will unconditionally attempt to protect. Along with the protection planning, a well-defined process must to be established to ensure that every other enterprise function is fully restored within requisite parameters. This is comparable to the "Plan-Purpose-Scope-Relationship" recommendations embodied in NIST SP 800-34—revision 1. Given that resources are frequently limited, the assets that are likely to fall under the recovery function versus protection will far exceed those that are being explicitly assured. Therefore, the scope of the Recover stage is most likely to exceed the protection operation by many magnitudes of effort.

7. *Evolve*: The one thing that is certain about the threatscape is that it is constantly evolving. New threats appear every day and old threats mutate into new forms that have not been accounted for. Therefore, the organization needs to dynamically adjust its actual cyber-resilient architecture based on direct feedback from the operation and the lessons learned from that. This is a continuous and ongoing stage that is different from the control deployment

and the recovery processes. It is also much more cyclical in both planning and execution. In that respect, it is comparable to the continuing execution of the process that is outlined in the NIST Cyber Security Framework (NIST-CSF, 2014a). Because it is meant to be continuous, once this stage is attained the process of managing the evolution of the protection scheme will characterize the actual cyber resilience effort for that organization.

Practically, the development and maintenance of a fully cyber-resilient organization involves strategic planning. In case of cyber resilience, the aim of strategic planning is to ensure that all the people, processes, and technologies that comprise the organization's substantive resilience response work together to achieve the purpose of cyber resilience. That requires the business to undertake a formal, organization-wide process to define, integrate, and maintain proper alignment between its business purposes and the assurance of critical functionality.

In this respect, the plan provides a definition of the assurance practices for each critical component as well as a statement of the resource commitments that will be required in order to assure the successful operation of that function. In addition to resources, the plan also specifies the administrative model that will be used to assure proper management oversight.

The outcome of the resilience planning process is an integrated set of organization-wide security controls. It probably goes without saying that in order to be effective, this complex and diverse set of actions has to be coordinated. That coordination normally requires a company-wide process. The process itself has to be capable of ensuring the continuous protection of critical functions and assets.

Effective coordination of the cyber resilience function starts with the creation of an enterprise-wide process whose sole purpose is to integrate the domains of business and information technology resilience into a single unified protection concept.

The actual steps that are taken to implement such a system typically fall into three phases. The first phase involves the establishment of a relevant set of security goals for the resilience function as a whole. In the second phase, the organization establishes the requisite level of assurance for the control set. Factors that might enter into decisions about the level of assurance include the level of criticality and the commensurate degree of threat.

Keywords

Access control: Authentication, authorization, and assurance of legitimacy of an access request

Architecture: The designs and implementations of an underlying framework of processes

Best practice: A set of lessons learned, validated for successful execution of a given task

Baseline: The collection of a set of objects all commonly related by application or purpose

Control design: Specification of behaviors of a protection measure or measures

Control evaluation: Formal testing or reviews of a control or control set to confirm correctness

Control performance: The operational results of control operation within a given environment

Controls: A discrete set of human or electronic behaviors, set to produce a given outcome

Critical asset: A function or object that is so central to an operation that it cannot be lost

Cyber resilience: Assurance of the survival and continued operation of critical assets

Cybersecurity: Assurance of confidentiality, integrity, and availability of information

Resilience: The ability of a given entity to continue to function under any adverse condition

Reliability: Proven capability to perform a designated purpose over time

Resilience management: Formal oversight and control of resilience actions of an organization

Strategic planning: The process of developing long-term plans of action, aimed at furthering and enhancing organizational goals

References

Bradford, C. (2017), Disaster recovery metrics: What they are and how to use them, *Recovery Zone*, www.storagecraft.com/blog/disaster-recovery-metrics-use/. Accessed March 2017.

Brasso, B. (2016), Cyber attacks against critical infrastructure are no longer just theories, [online], *Fire-Eye*, www.fireeye.com/blog/executive-perspective/2016/04/cyber_attacks_agains.html.

Clausewitz, C. (1989). *On War*, Howard, M. E. and Paret, P. (Editor and Translator), Princeton University Press.

Conklin, W.A., Shoemaker, D., and Kohnke, A. (2017), *Cyber Resilience: Rethinking Cybersecurity Strategy to Build a Cyber Resilient Architecture*, 12th International Conference on Cyber Warfare and Security, Dayton, OH.

Eisenhauer, J., Donnelly, P., Ellis, M., and O'Brien, M. (2006), *Roadmap to Secure Control Systems in the Energy Sector*, Energetics Incorporated, Sponsored by the U.S. Department of Energy and the U.S. Department of Homeland Security, January 2006.

Ernst and Young (2014), Achieving resilience in the cyber ecosystem, [online], *Ernst and Young*, www.ey.com/Publication/vwLUAssets/cyber_ecosystem/$FILE/EY.

Gamer, N. (2015). A decade of breaches: Myths versus facts. Trend Micro, Simply Security. Retrieved from https://blog.trendmicro.com/a-decade-of-breaches-myths-versus-facts/accessed.

Hatchimonji, G. (2013), Survey results reveal both IT pros' greatest fears and apparent needs, *CSO* On-line, Sep 18, 2013, www.csoonline.com/article/2133933/strategic-planning-erm/survey-results-reveal-both-it-pros--greatest-fears-and-apparent-needs.html. Accessed January 2017.

Howard, M., Paret, P., and von Clausewitz, C. (1976), *On War*, Princeton University Press, Princeton, NJ.

Hulme, G.V. (2013), *State of the CSO in 2013 Shows Improved Outlook*. Retrieved from www.csoonline.com/article/2134240/employee-protection/state-of-the-cso-in-2013-shows-an-improved-outlook.html Dec 2017.

Lois, J.E. (2015), It can happen to you: Know the anatomy of a cyber intrusion, *Navy Cyber Defense Operations Command (NCDOC)*, Story Number: NNS151019-05, Release Oct 19, 2015.

Maynor and Graham, R. (2006), SCADA security and terrorism: We're not crying wolf, *X-Force, Black Hat*, file:///C:/Users/dansh/AppData/Local/Temp/BH-Fed-06-Maynor-Graham-up-1.pdf. Accessed Mar 2017.

National Institute of Standards and Technology. (2014a), *Cyber Security Framework*, NIST, Gaithersburg, MD.

National Institute of Standards and Technology. (2014b), *NIST Special Publication 800-53A Revision-4, Assessing Security and Privacy Controls in Federal Information Systems and Organizations Building Effective Assessment Plans*, National Institute of Standards and Technology, Gaithersburg, MD.

National Institute of Standards and Technology. (2014c), *Risk Management Framework*, NIST, Gaithersburg, MD.

National Geographic Channel, (2017), American blackout. Accessed Mar 14, 2017.

OAS. (2015), Report on cybersecurity and critical infrastructure in the Americas, Organization of American states, [online] Trend Micro Incorporated, www.trendmicro.com/cloud-content/us/pdfs/security-intelligence/reports/critical-infrastructures-west-hemisphere.pdf.

Privacy Rights Clearinghouse. (2015), *A Chronology of Data Breaches*, PRC, San Diego, CA.

Privacy Rights Clearinghouse. (2016). *Privacy Rights Clearinghouse, "A Chronology of Data Breaches,"* PRC, San Diego, CA.

Saltzer, J . H., and Schroeder, M. D. (1974). The protection of information in computer systems. *Communications of the ACM*, 17(7): 338–402.

Smith, A. (2014), China could shut down U.S. power grid with cyber attack, says NSA chief, *Newsweek*, Nov 21, 2014.

Symantec. (2014), A manifesto for cyber resilience, [online], *Symantec*, www.symantec.com/content/en/us/enterprise/other_resources/b-a-manifesto-for-cyber-resilience.pdf.

Trend Micro. (2015), Report on cybersecurity and critical infrastructure in the Americas, Organization of American States, Trend Micro Incorporated.

US-CERT. (2016), Cyber resilience review (CRR), [online], Department of Homeland Security, www.us-cert.gov/sites/default/files/c3vp/crr-fact-sheet.pdf.

Verizon. (2014), Data breach investigations report, Verizon Corporation, 2014.

Wagner, D. and Schweitzer, B. (2017), The growing threat of cyber-attacks on critical infrastructure, *THE BLOG*, May 25, 2017, www.huffingtonpost.com/daniel-wagner/the-growing-threat-of-cyb_b_10114374.html accessed July, 2017.

White House, The. Presidential Policy Directive 21: Critical Infrastructure Security and Resilience (PPD-21), Feb 12, 2013.

Chapter 2

Asset Identification and Classification

Following this chapter, the reader will understand

1. the relationship between standard Information Technology Asset Management practices and cyber resilience;
2. the importance of and tasks involved in asset management planning;
3. the tasks necessary to identify and classify people, information, technology, software, and facility assets;
4. how to use configuration management to establish an asset baseline and establish relationships between assets and critical services;
5. how to use configuration management to effectively manage changes to services and the assets they support in order to sustain cyber resilience;

Laying the Groundwork

It should be clear from the discussions in the first chapter that a thorough understanding of the assets an organization is protecting through cyber resilience is at the core of providing the protection necessary to maintain organizational operation of vital business functions during and after cyberattack. The problem is that many organizations don't know where to begin, much less have a clear vision of what assets must remain resilient. Regrettably, many organizational efforts in identifying and classifying assets result in several haphazard attempts to "bean count" what is believed to be vital to success without any realistic justification for decisions that have been made. It is sad to say that the reason for this chaotic approach is because there tends to be a misconception that what is being protected is strictly within the realm of the Information and Communication Technology (ICT) function of the organization and only relates to the notion of confidentiality, integrity, and availability of information assets. In reality, the ability to appropriately identify and classify organizational assets for cyber resilience requires a considerable amount of enterprise level analysis and subsequent process development. Stated differently, a methodological approach must be established in which the organization must first understand its core mission and strategic objectives. Further, every service achieving that mission and objective (internally and externally) must be identified. Next, the critical assets, either produced or utilized within those services, must be appropriately exposed for consideration of their resilience requirements. Based on those requirements the assets can be appropriately classified.

A common misconception, stemming from the nature of the word *cyber*, is that all cyber-resilient assets identified by the organization must be information based. While information assets cannot be disregarded (and perhaps the most vulnerable), the scope of critical assets reliant on resilience mechanisms extends well beyond just information resources and includes: hardware, software, human resources (HR), databases, communication channels, customers, suppliers, and business processes that serve as a focal point of the organization's critical infrastructure.

In this chapter, we will present a framework for asset identification and classification that is the first of many organizational processes aimed at providing an appropriate level of cyber resilience. Theoretically, what is under discussion here, is not new. For many years, organizations have included some form of asset management into their IT (information technology) function. We will begin with a general discussion of asset management and then take that conceptual understanding into the depths of applying it in such a way that the organization can build it into the maturation of the larger scope of a cyber resilience program.

Putting Asset Management into Cyber Resilience Context

Once known primarily as a source for inventory management of physical IT assets, IT Asset Management (ITAM) has grown to support the documentation and management of an organization's IT assets (to include hardware, software, communications, and people resources) as they progress through the asset life cycle: from acquisition to installation and configuration, usage, updating and reassignment through to decommissioning and disposal. Without a systematic approach in place to manage these (and other) processes, things can either go awry or get left out altogether, thereby exposing the organization's financial loss (of assets that cannot be located) or security risks (if software upgrades or security updates have been missed or unauthorized software has been installed). The implementation of an ITAM program not only gives the organization a clear view of the location, inventory details, and performance of assets but also helps them to record relevant financial information and keep on top of regulations surrounding licensing and contracts management and security.

ITAM has become such a vital part of providing a solid mechanism for effective critical infrastructure that the National Institute of Standards and Technology (NIST) has included it as one of the five categories of the Identify Function of the Framework for Improving Critical Infrastructure Cybersecurity (ICSF).

Given the contextual similarities of cybersecurity and cyber resilience, it might be wise to use the ICSF definition of asset management as a basis for further discussion. NIST defines asset management as: "The data, personnel, devices, systems, and facilities that enable the organization to achieve business purposes are identified and managed consistent with their relative importance to business objectives and the organization's risk strategy" (National Institute of Standards and Technology, 2014).

What that definition emphasizes is that asset management is a set of IT processes designed to manage the life cycle and inventory of technology assets. It provides value to the organizations by lowering ICT costs, reducing IT risks, and improving productivity through proper and predefined management practices. Asset management has existed only as a formal set of IT processes for about a decade, which is immature in comparison to typical IT life cycle processes. Although it may not be completely obvious where asset management and cyber resilience come together. After all, cyber resilience is a complex activity involving highly skilled IT engineers, architects, and strategists charged with defending against everything from run-of-the-mill spam and phishing to multinational terrorism cells insistent on causing instability in financial markets through computer-based terrorism and everything in between.

Asset management has many goals, including maximizing the value of an organization's investment in IT. One common approach to achieving this goal is through understanding the IT needs of the organization and then establishing standards that serve to facilitate those needs. That in turn leads to the justification of asset types and, more often than not, the reduction of asset types. For example, organizations can see a significant reduction in the number of software applications through an application-justification process; this involves defining which types of applications meet the predefined guidelines that support the organization's IT objectives and working to remove the applications that do not meet the guidelines. With the elimination of each application comes increased resilience because that is one less application to harden, patch, monitor, and audit, that must be included in the asset baseline.

Another benefit of the asset management is the increased understanding of which individuals within the organization need each of the assets in order to perform their roles within their business environment. Organizations that practice access management, for example, understand who has access to sensitive data and user permissions can be more logically restricted based on need, in some cases even serving as the basis of or logic check for privilege management systems.

Cyber Resilience Asset Management Processes	
Asset Management Planning	Obtain support from higher management to ensure that appropriate funding exists, staffing is available, and asset management activities are performed.
Identify Assets	Identify existing assets based on predetermined asset definitions, then provide classification of each asset relative to type, sensitivity, location, and owner.
Document the Assets	Prepare an asset inventory based on identified assets and match the mission critical services to the assets that support them.
Manage the Assets	Provide the management necessary to correlate and coordinate change and inventory of the assets and improvement of the asset management process.

Figure 2.1 Cyber resilience asset management processes.

To ensure all assets, applicable if the organization's resilience priorities are properly identified and classified to the extent that vulnerabilities can be exposed, a framework of individual process groups can be used to ensure that an organization implements asset management to the degree that ensures organizational assets can truly become cyber resilient. Each process group is further divided into activities and tasks that assist the organization in measuring the maturity of their implementation. Figure 2.1 provides a summary of each process group and each is explained in detail in its own subsequent section of this chapter.

Asset Management Planning

As indicated in the previous section, the main objective of asset management is to give an organization a snapshot of all the assets within its critical infrastructure at any moment in time and allow the organization the capability to make decisions regarding the resilience requirements of each asset. Consistent with the best practices of other functions of IT, developing and following a plan for asset management is an essential by-product in order to establish and maintain efficiency of each process and activity. When organizations implement the planning process at the outset they are ensured that support from senior level management is obtained, maintaining assurance that the process is funded, having preliminary insight into what will be necessary for the process to be properly staffed, and understanding the necessities for performing each activity and task at a level necessary to achieve cyber resilience requirements.

Equally important to obtaining support from senior management, asset management planning also includes understanding the mission of the organization and then identifying all the mission-critical services performed or provided. Further, each service is prioritized according to its potential to disrupt operations, should that service fail entirely and be limited in functionality. With the appropriate knowledge of services, the organization can manage its resources to appropriately protect its assets and sustain operability.

Lastly, perhaps one of the most vital objectives of planning for asset management is that it requires the organization to establish a common definition of what constitutes an asset type within its infrastructure. This determination should not be the result of a decision made by any one individual within the organization. Rather, a methodological approach must be taken to justifying assets via groups of individuals (from senior management through lower level managers), in consideration with organizational culture and each tangible and intangible source of value to the achievement of the established mission, strategies, and organizational objectives.

Obtain Support for Asset Planning

Effective asset management involves the implementation of a fully integrated set of organization-wide management procedures and technical elements. That complex set of components normally requires a company-wide process to manage it. The process itself has to be capable of ensuring a productive relationship between the business goals of the organization and the asset management process. The term that describes this optimum state is alignment.

Because it has company-wide implications, the responsibility for ensuring proper alignment and coordinating the overall process has to have the direct support of the Chief Executive Officer (CEO). The Chief Information Officer (CIO) and Chief Information Security Officer (CISO) represent the executive leadership of the technology function, so one or both of them have to be accountable for maintaining the productiveness and efficiency of the asset management process and should be able to provide quality feedback to the CEO in order for leadership of the process to be sustained. Since diverse perspectives are involved in the actual planning, senior management support must be in place to also ensure that the people needed to define and maintain an effective alignment can work well together. Likewise, a top-down approach is not only useful, but necessary for the asset management process to meet the resilience objectives of the organization. That support becomes vital within the process in terms of funding, oversight, and staffing. Additionally, considering the implications of supply chains, a top-down approach enables consistency for implementing asset management across organizational boundaries.

Much like the implementation of life cycle processes (which requires "across the board" management oversight), the amount of management support required within the asset management process largely depends upon where the assets are within the organization. Generally, two scenarios could exist as shown in Figure 2.2.

The first considers an asset management plan that addresses the entire organization. In such a case, senior management support is required in addition to a

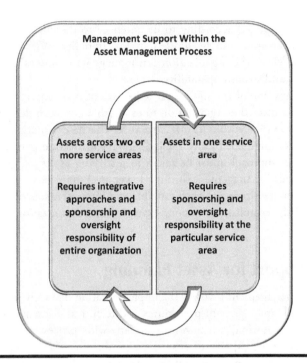

Figure 2.2   Management support requirements.

larger extent of oversight responsibility that is carried by the need for integrative approaches to managing the assets across two or more service areas. In a second scenario that may implement asset management at the service level, senior management support is still a major element of process success, but a larger extent of sponsorship generally comes from management responsible for that particular service area. For example, consider a large insurance organization that provides insurance (home, car, and life), variable life annuities, structured settlements, commercial mortgages and securities, sovereign debt, and other business support services. Asset management could be implemented for each of those services individually with sponsorship coming from each business unit. Alternatively, senior management of the insurance company could require a common asset management process for all business services combined.

Service Identification

Identifying assets within an organization can be a daunting task, so the organization should first identify the services it provides. It can then identify assets by the services they support and divide the body of assets into manageable pieces. However, before services can be accurately identified, a holistic understanding of the organization must be realized. This is done through a review of the organization's strategic objectives and the creation of a formalized set of critical success factors (CSFs).

Stated simply, strategic objectives are a set of targets that the organization identifies as a means by which it will be able to accomplish its mission, vision, values, and purpose. Communicated effectively, the information needed to identify these objectives should be available in organizational literature such as staff handbooks, annual reports, or the corporate website. In some large organizations, this information may also exist at, and be unique for, each organizational unit or service area rather than at just the enterprise at large. The main point of collecting this information is to serve as an organizational road map for performance so that everyone within the organization is moving in the same direction.

The underlying goal of cyber resilience is to provide the capability for the organization to always reach its strategic objectives. Management practices that aim toward cyber resilience must be consistently focused on enabling the achievement of those objectives that allow an organization to meet its established mission, vision, values, and purpose by addressing all the potential disruptions that can limit their achievement. Having originated from the organization's strategic planning process, strategic objectives should be appropriately identified within the organization's strategic plan.

Based on a clear understanding of strategic objectives, the organization can engage in a data-driven decision making process aimed at establishing a set of CSFs. CSFs are the limited number of service areas within the organization that must continue to perform their operations indefinitely in order to meet the identified strategic objectives. Because of the close ties to established objectives, these service areas require a high amount of managerial implicit focus. Once the CSFs have been identified, defined, and communicated, they represent the core set of criteria by which the organization should align its activities. Stated differently, CSFs provide the organization with a set of performance indicators that all cyber resilience initiatives must provide support by establishing alignment between the policy-making and the operational levels of the organization.

Once an acceptable set of CSFs has been established, services that support those factors can be identified. Services are those activities that the organization performs as a contribution to its customer or in the production of a product. Because these services directly support the achievement of strategic objectives, they must be protected and sustained in order to minimize disruption. Organizations without a plan to keep these services operating normally risk the inability to meet strategic objectives. The most valuable resources that organizations use in identifying services linked to CSFs include:

- strategic plans
- business plans
- contracts
- and operational business processes

Services are provided by an organization through internal processes or external affiliations or both. As such, organizations should prioritize the understanding of how each service contributes to the CSFs, and how the combination of internal and external activities that are performed within that service are interconnected.

As the organization facilitates the activity of service identification, the nature of the service, alignment to strategic objectives, and resilience requirements must be appropriately documented. The mechanism used will vary by organization. However, some approaches include a simple spreadsheet, a more extensive configuration-controlled document, or a database. Essentially, the documentation is merely a service catalog with the information needed for implementation of resilience measures. Information contained within the service catalog is collected as each process of cyber resilience is implemented; it includes but is not limited to

- service inputs and outputs;
- assets associated with or used by the service;
- owners and stakeholders of the service;
- related services and business processes;
- resilience requirements;
- and service level requirements including responsibilities, availability, hours and exceptions, response times, and consequence of failure.

Each service must be appropriately logged and entries subjected to configuration management (CM) policy (discussed later in this chapter) set forth by the organization. In doing so, subsequent changes to any service will be considered with regard to the affect that change has on the resilience of affected organizational assets. Figure 2.3 shows the elements involved in identifying services.

Service Prioritization

Once all organizational services have been identified, each is then prioritized in an effort to "single out" high-value services. In essence, high-value services are those that are critical to the success of the organization's mission, where their failure may present the likelihood that the organization's strategic objectives may not be met. Another term often used for this activity is "affinity analysis." One of the first steps in service prioritization is for the organization to decide how it will assign value to each identified service. When assigning value and prioritizing services, the organization should consider the following:

- results from previously performed business impact analysis (BIA)
- details contained within business continuity plans
- risks to services identified during risk assessment activities
- and organizational strategic objectives

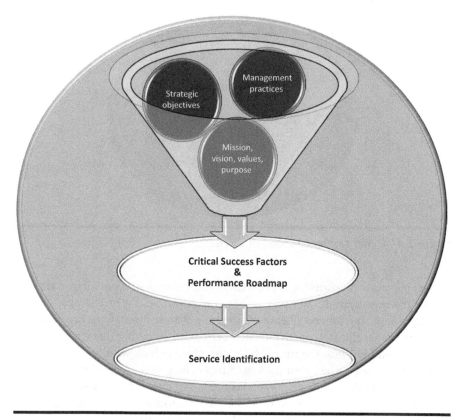

Figure 2.3   Elements of service identification.

It is important to note that there is no definitive method for how to proceed through service prioritization. The reason is because of the unique nature of each organization, the services it provides, interdependencies between services and supply chain partners, and relationships with customers. Nevertheless, regardless of the methodology used, once the organization has prioritized its services and appropriately labeled those it considers to be of high value, it should document this information in the service catalog created or updated during service identification.

By properly prioritizing services, an organization can proportionately allocate budget and resources for resilience activities with the services that matter most, such as identifying assets.

Establish a Definition of Assets

Once the organization has a clear understanding of the services it provides (mission, vision, and objectives) and has appropriately prioritized those services, the final task

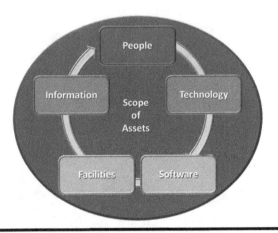

Figure 2.4   Cyber resilience scope of assets.

of asset management planning is to establish a common definition of assets existing within the organization. Without a common definition, every attempt to identify assets requiring cyber resilience will become chaotic through the organization. Understand that this step doesn't identify each asset. In taking that leap, the organization would fall into the rut of "bean counting" (which we alluded to earlier). Rather, there needs to be a clear understanding of how the organization perceives as being characteristic of each asset type. As much as a clear set of strategic goals and objectives must be established in order for the organization to be managed within the capacity of achieving those goals and objectives, everyone involved in the successful performance of prioritized services must have a consistent interpretation of the assets created or used by the service. Unlike the practices of cybersecurity, which is generally based on protecting information, cyber resilience expands the scope of asset definition to include (Figure 2.4):

- people
- technology
- software
- facilities
- information

People

Most cyber resilience literature suggest that people assets are merely the vital staff who operate and monitor the organization's services. While that premise does carry some truth, we suggest that the scope of the definition be expanded to state something as abstract as: "People are those individuals (internal or external to the organization) who are vital to the operation and performance of identified services."

By using the later definition, we are not limiting ourselves to just the staff that perform the service. Rather, we can also include customers (i.e., project sponsors), supply chain partners, and other service specific stakeholders.

When defining people assets, the organization should consider the vital role required for the successful operation of a service rather than just focusing on the actual individual in that role. We suggest that each role contain a defined list of the functions or responsibilities required in the performance of that role.

Technology

Technology assets include hardware, firmware, and any guided or unguided media that provide interconnections between each technology component. Technology is often layered throughout the organization. Some are specific to a service (like a patient monitoring system in a health care institution), while other technologies are part of a larger enterprise architecture that shares the assets depending upon the service provided. Still, other technologies span across organizational boundaries and affect the performance of services within supply chain partners.

Software

The nature of software can be very large in scope and fairly daunting to define, in terms of asset type. Organizations must consider the system level software used to support the functionality of each information subsystem, the Commercial off the Shelf (COTS) and proprietary software, and the software that extend beyond organizational boundaries into supply chair organizations. It is recommended that the definition of this asset type be kept simple. So, simple in fact, that an organization can begin their definition by looking at software as falling into one of three major categories: applications, tools, and infrastructure. Further definition of software within each can be performed by organizations as cyber resilience requirements warrant.

Applications are software built to help a user do some valuable activity, like perform payroll activities, check a bank balance, or edit digital images. While the user must learn how to execute its functionality, an application is generally apt to satisfy functional requirements without further development. However, some applications (such as COTS software) are customizable. With that capability, each application could work differently for each user or organization, but they are still focused on solving problems or delivering value related to some specific functional domain.

Often, tools tend to be defined only by their natures to add on value to system software. However, a more generic definition can be established. Tools are software that are "general purpose," but with a specific functional value. They are designed so that users can execute functionality that allows for the configuration of applications for individuals or workgroups that could consist of users in a specific service area across the organization or further to the extent of supply chain partners. Tools express their own user experiences, but are not always immediately valuable without some form of configuration. Moreover, they can range from simple Microsoft Excel spreadsheets to SharePoint communication resources to web content management systems like WordPress or large-scale enterprise resource planning (ERP) systems like SAP or PeopleSoft. Many business intelligence products such as expert systems, knowledge management, and decision support systems fall in this category.

Infrastructure software tend to have the least user interaction, beyond the configuration necessary by system administrators and other IT staff. This type of software is designed completely as a foundation for other software to be built upon. This would include any software whose primary interaction mode is through API (application programming interfaces) or CLI (command line interface) patterns. Products like databases, middleware, application servers, application frameworks, and operating systems fall in this category.

Facilities

Facility assets are any physical space an organization relies on when delivering or performing a service. Owned and controlled by the organization or under the control of external supply chain partners, facilities are the places where the services are executed and are often shared in such a way that one or more services may be dependent on them. Put differently, facility assets provide the means by which the people, technologies, software, and information assets are able to contribute to the delivery of organizational services and provide the protection capability for all those assets. Because of the close correlation between facilities and other assets, it follows that resilience planning for facilities would be tightly integrated with the planning of other asset types.

Information

Finally, information assets are any information or data, on any media, required for the successful operation of an organizational service. To that extent, an information asset can also be the intended input, output, or other results of a service. Such information can range from a bit or byte, a text or binary, or something as simple as a word-processed document, to the collection information stored in a database. Because of the degree from which information is generated, stored, and utilized within each service, the organization must determine the level of detail with which it wants to define its information assets. That level of detail must be appropriately disseminated to all organizational parties charged with the responsibility of creating a resilience definition that can be applied across the organization.

Identify the Assets

When asset management planning is complete, the organization should have support for asset management from senior executives, a thorough understanding of how services provided by the organization map to the identified factors that are perceived as critical for achieving strategic goals and objectives, and a consistent definition of each asset type (as they relate to that organization's mission and objectives). The second major activity of cyber resilience asset management begins by identifying those assets that are critical to the continuity of the operations of the organization and therefore fall within the scope of resilience requirements.

As each asset is identified, they must be mapped to the service they support. By creating an association between assets and services, the organization is able to better understand where critical dependencies exist, afford them the capability to validate resilience requirements, and extend the capability of developing and implementing a defined set of resilience strategies.

The tasks involved in identifying assets include

1. assigning responsibility for identifying assets that support critical services;
2. identifying people assets;
3. identifying and classifying information assets;
4. identifying technology assets;
5. identifying software assets;
6. identifying facility assets.

Assign Responsibility for Identifying Assets that Support Critical Services

The vital first step in asset identification is the task of identifying key stakeholders for each critical service. While it is beneficial to have identified each individual service, knowing the stakeholders of that service is essential to the understanding of how the organization approaches the larger scope of service design. Likewise, these stakeholders often support multiple services within the organization and can offer unique insights on the assets involved in supporting them (individually and in combination). Generally, stakeholders of organizations serve in work responsibilities, which could be through a formal assignment process in large complex organizations or in smaller organizations, informally assigned and based on traditionally performed work responsibilities.

The two most critical stakeholders to identify are asset owners and asset custodians. Asset owners can be individuals or organizational service areas, internal or external to the organization. Their primary responsibility to any particular asset is to maintain its viability, productivity, and resilience. For example, a people asset such as what the organization might label as "nonexempt staff" may be owned by the HR department or the HR Manager. It is the owner's responsibility to ensure that the appropriate levels of confidentiality, training, and availability requirements are defined and satisfied to keep the asset productive and viable for use in their appropriate services within the organization.

Asset custodians can also be individuals or organizational functional units, internal or external to the organization. They are responsible for the implementation of management controls to satisfy the resilience requirements of their associated asset(s) while they are in their care. For example, a subset of the nonexempt staff in the above example may physically work within the customer service department. Fundamentally, the Customer Service department takes custodial control of that subset of staff (from the perspective that they are considered as a high-value asset within the organization). The Customer Service department must commit to taking actions commensurate with satisfying the owner's (in this case the HR department) requirements to protect and sustain the asset. In practical terms, however, it is the responsibility of the owners to ensure that their assets are properly protected and sustained, regardless of the control mechanisms put into place by the custodians.

The relationship between custodians and owners is not always straightforward, however many challenges may occur in ensuring that the asset's resilience requirements are met. In many cases, an asset does not have just one asset owner. As such, custodians of those assets must resolve conflicting requirements obtained from each owner. Likewise, the custodian may be an individual or supply chain partner external to the organization. In these cases, the asset owner(s) must be equipped to communicate the resilience requirements of their assets to those external custodians, and must implement additional controls aimed at monitoring the fulfillment of those requirements.

In order to see the relationships between stakeholders and each asset type, the organization should develop a Responsible, Accountable, Consulted, Informed (RACI) matrix similar to Figure 2.5. In the matrix, the role/asset type relationship is based on four distinct categories. A "responsible" role manages the asset on a day-to-day basis. An "accountable" role has to answer to higher level management when issues arise with that asset. A "consulted" role has specific knowledge of the asset or

Role	Position	Asset Type: People	Asset Type: Technology	Asset Type: Facilities	Asset Type: Information
Executive Sponsor	Executive VP/ VP/Director	Accountable	Informed	Informed	Informed
Service Delivery Manager (business process owner)	Functional Manager/IT Manager	Responsible	Accountable	Consulted	Accountable

Figure 2.5   RACI matrix of stakeholders and assets.

relies on that asset and needs to provide input. And an "informed" role would need to be updated on the activities surrounding that asset.

Identify People Assets

An organization's ability to achieve mission and objectives relies largely on the dependencies between those organizational goals and objectives, services that are critical for organizational success and continuity and the assets required to perform those services. If any of those assets are missing, inoperable, or lack performance necessary to perform the services (as a result of disruptive events, threat exploitation, or other issues such as natural disaster) and prevents the service from critical operation, it can present a consequence of nonexistent mission assurance and failure to achieve organizational goals and objectives. Therefore, using the asset definitions previously developed, the organization must ensure the cyber resilience of the assets associated with those critical services.

Because assets derive their value and importance through their association with services, the organization must identify those assets that, in their absence, critical services cannot operate without. In performing the asset identification task, the organization should look at assets from a holistic perspective. In other words, the organization should ask itself which assets can we not live without. Later, those assets will be associated with appropriate services. This top-down approach ensures that all critical assets are identified by considering them twice; once at the enterprise level and then again at the service level. Additionally, duplication of assets that are associated with multiple services is minimized. This approach also provides structure and guidance for later developing an inventory of assets from which resilience requirements have to be established and satisfied.

Without a doubt, organizations rely heavily on the experiences, talents, and skills of internal and external people to support critical services. Management, from the senior level all the way down to middle and line levels, admits that employees (whether direct hire or contractors) working for the organization are key to its success. They are responsible for executing the organization's mission and are a key component in its ability to adapt to change or disruption and remain resilient. By focusing on employee work roles, the organization is able to capture the critical knowledge, skills, and abilities required for each position within critical service areas.

Once employed, people become a vital part of the organization as a whole and the service or services where their job responsibilities and accountabilities are managed. In consultation with management of each service for which the employee skills are utilized, the HR department is considered the "owner" of these resources and must work in collaboration with management of the appropriate service area in establishing authority and accountability for work assignments and training, deployment, and performance.

Role	Level of Experience Required	Educational Qualification	Required Knowledge	Required Skills	Certification(s)
Accounting Manager	8-10 years	B.S. in Accounting, Masters Preferred	Accounting Management	Excellent written, oral communication, Decision-making, Organizational skills	CPA
Project Manager	5-7 years	BS in IT, CIS, or CS, Masters Preferred	Critical Path Analysis, Cost analysis, Budgeting, Risk management, Procurement management, Stakeholder management, Quality management, Scope & Time management, Agile methodology	MS Project, MS Visio, SharePoint, Excellent written, oral communication, Life cycle management, Leadership, Negotiation, Team building, Mentoring	PMP; IPMA: CPM, CPD; RPM; Australian AIPM, Japan P2M: PMR PMA

Figure 2.6   Identification of people assets.

As people assets are identified, information about those assets will need to be collected and documented in a people asset register. Such a registry closely resembles a document of qualification standards that organizations often use in the hiring process. Figure 2.6 shows examples of the types of information that should be documented when identifying people assets.

Identify and Classify Information Assets

With a clear definition of information assets established during asset management planning, the organization must identify each trace of information that is vital to the success of the organization. Each identified information asset must be capable of being mapped to associated critical services in a later activity. Further, the information asset type is unique in terms of the emphasis placed on the tasks performed in applying the appropriate classification based on the amount of risk it imposes to the underlying resilience of the organization.

Information Asset Identification

The collected information should not only include the technical documentation involved in supporting the service, but also the actual information contained within the databases utilized by the service. Every organization will have a slightly different approach to identifying information assets. However, regardless of the methodology, those information assets to consider include but are not limited to (Figure 2.7)

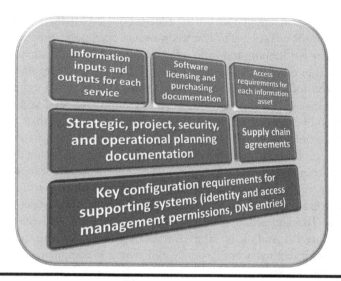

Figure 2.7 Information asset identification.

- information inputs and outputs for each service;
- strategic, project, security, and operational planning documentation;
- inbound and outbound information flows between customers and supply chain partners;
- key configuration requirements for supporting systems (e.g., identity and access management permissions, domain name system (DNS) entries);
- software licensing and purchasing documentation;
- supply chain agreements and the people authorized to communicate with the suppliers;
- and access requirements for each information asset.

A truly resilient organization must be able to find, open, use, understand, and trust the information in order for it to be valuable to the organization. It is important to note that during this activity part of the focus is on mapping the information to critical services. Likewise, identification of organizational information assets would not be possible without the assistance of each information owner and information custodian. It is the responsibility of the information owner and information custodian of each service to ensure that the information generated from or in support of their service is appropriately inventoried, cataloged, and archived in accordance with the organization's security policies and any regulatory or legal compliance requirements. Information within an organization may be bound by retention requirements that must be constantly monitored to ensure that only the essential information required for the execution of that critical service is maintained for the specified time line. Keeping too much or too little information introduces additional risk to the organization that the underlying principles of cyber resilience attempt to avoid. Keeping too much information also induces limitations on resources and adds unanticipated cost to the overall operation structure of the organization. Likewise, keeping too little information may open the organization to a variety of fines or other punitive measures defined by applicable regulatory regimes.

The data custodian must also ensure that appropriate controls are enacted to protect the information assets of the critical service. Enterprise-level and service-level controls should be defined, implemented, and assessed to ensure that the appropriate level of confidentiality, integrity, and availability of the information assets are maintained. It is the management's responsibility, therefore, to ensure that the organization's policies clearly define the appropriate implementation of those controls.

It is also necessary that the information owner and information custodian constantly know exactly who has access to the types of information under the custodian's purview. This authoritative measure should be reviewed on defined scheduled basis by an Internal Auditor or other third-party auditing organization to ensure compliance with defined security policy and that it is compliant with legal and regulatory requirements.

Information Asset Classification

Once the task of holistic organizational information asset identification is complete, the organization should move on to the activity of classifying each information asset. This activity largely entails selecting provisional impact values for each informational asset. Provisional impact values relate to the degree in which that information asset requires one of the three underlying objectives of information security: confidentiality, integrity, and availability. If the identification activity has been performed correctly up to this point, each information asset can be assigned a provisional impact value of low, moderate, or high. Confidentiality can also have an impact value of not applicable when the information asset contains public information. At the conclusion of this activity, each identified information asset should have been preliminarily documented and categorized based on the following syntax:

$$
SC_{\text{information type}} = \left\{ \left(\text{confidentiality, impact} \right), \left(\text{integrity, impact} \right), \left(\text{availability, impact} \right) \right\}
$$

where "SC" is indicative that what is being documented is a security category, "information type" is the identified information asset, and impact is either low, moderate, or high as a rating mechanism for that asset's impact on the organization, based on each information security objectives of confidentiality, integrity, and availability. The recommended provisional level is then justified for how that impact level was determined for each of the three security objectives.

Many organizations question the need for recording potential impact types. Too often, management claims that it is a waste of time. However, it is important to remember that, in a subsequent activity, the documented analysis performed in this task will be used to create an asset inventory for each critical service. That document will be read and interpreted by different groups of people for a variety of purposes. It is necessary to continuously provide documentation to the extent that the information is available for the right people, in the right place, and at the right time.

Provisional impact levels are not meant to be permanent representations of each information asset. As the organization continues to identify threats and vulnerabilities that affect the cyber resilience standards, new information systems will be added and new information assets identified. Thus, organizations must continuously evaluate the assigned impact levels of existing information assets and follow the proper protocols for the assignment of initial provisional impact levels for new information assets affected by known threats and vulnerabilities.

Identify Technology Assets

Consistent with the identification of each of the other asset types, proper identification of technology assets must put the organization into a position that the technology can be further matched to appropriate critical services. This evaluation of technology will often focus on the IT system/subsystem layer of the enterprise infrastructure, but the underlying components and their interdependencies must also be carefully considered.

The technology assets should include all the components involved in delivering or supporting service(s) throughout the organization. Identifying the systems on which an organization's services rely is essential to understanding the overall resilience posture.

The architecture of organizational services and the technology assets supporting them can be illustrated using a layered approach. Organizations should categorize technology based on their operational contributions and traditional IT contributions. Extending downward from each of these two major contributive classifications, the specific technology components that belong to each of these major groups can be clearly identified. For example, a public works department of a large municipality could make use of operational technology such as remote terminal units that depend upon telemetry and programmable logic units that make use of a data acquisition server containing a Human Machine Interface (HMI). That same municipal department could also utilize IT with application software requiring data services (to be considered in the identification of software and information assets, accordingly) and computing and storage platforms that make use of networks and access management systems. The organization may not identify the lower-level assets at first, but the layered view of the service and asset architecture makes it obvious that these assets are important to the function of the top-layer assets.

Each of the technology components should be included in an enterprise CM database and managed through formal change control processes (discussed in a later section of this chapter).

Identify Software Assets

Dating back to the 1960s, 1970s, and 1980s when computers were primarily used for business data processing, there has always been a need for some form of software inventory and tracking capabilities. During that "mainframe era," keeping an inventory of software used within the organization was as easy as maintaining a defined CM program for the purpose of version control and using the records of that system to track the proprietary programs built within an organization's large data processing department.

Thirty years later, and in a time when cybersecurity and cyber resilience are main organizational priorities, the scope of software usage within organizations has increased to the extent that a plethora of different types of software are being used (often specialized to each service area) to effectively achieve organizational goals and objectives. Organizations are now faced with the reality that lack of a Software Asset Management (SAM) program will result in poor management of software resources, which will result in the inability to meet organizational objectives. The management of software assets has seen a large increase in demand and interest in its services in recent years, particularly with the increased emphasis on cyber-resilient organizations. There are a number of ways in which SAM can help an organization achieve its resilience requirements. However, in the same vein as other areas of ITAM, it takes substantial time and effort in combination with support from senior management.

There are several definitions of SAM that have evolved over the years. Microsoft, for example, describes it as "a best practice incorporating a set of proven processes and procedures for managing and optimizing your organization's IT assets. Implementing SAM protects your software investments and helps you recognize what you have, where it's running, and if your organization is using your assets efficiently" (Microsoft, n.d.). Nevertheless, the basis of SAM remains consistent. It relates to the processes, procedures, and people that are responsible for the management of software assets throughout their life cycle.

Through the use of specialized tools, at its core, an organization's SAM implementation must include seven vital policies or processes. Those policies and processes include considerations related to:

- Software use: highlights what users are permitted to do with any software installed on their machine.
- Software procurement: the correct process for purchasing new software instances.

- Software authorization and deployment: the process for new software requests and the correct approval/deployment process.
- HR management: what should happen with software when a new employee is hired if someone moves to a different service area or office, and also what happens to the software asset when an employee leaves the organization.
- Disaster recovery: in the consequence of a disaster, mission critical IT services are maintained until such time as the disaster is deemed concluded.
- Software recycling: ensures that the right methods are followed when redistributing or recycling a software license following someone leaving or an internal software license review.
- License compliance: ensures license compliancy is met, and that any non-compliance issues are addressed quickly and efficiently.

There are a number of other processes that an organization can implement, depending upon their cyber resilience requirements. However, the seven processes shown in Figure 2.8 are at the core for supporting the needs of a SAM program. Organizations shouldn't wait until resilience requirements are being considered. It is important for an organization to sustain a long-term SAM program that includes software as soon as it is built or procured. The success of SAM is primarily contingent on the successful implementation and sustainment of relevant processes, the ability to have the correct management in place to ensure that the processes are followed, abided by, and capable of improvement.

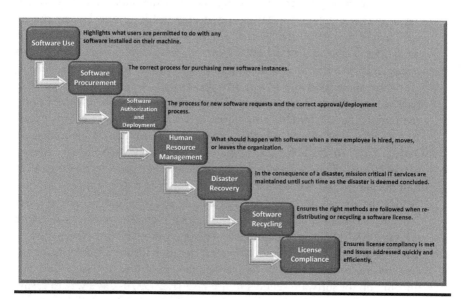

Figure 2.8   SAM implementation process.

Standardizing Software Asset Management

In 2006, the International Standards Organization in collaboration with the International Electrotechnical Commission (ISO/IEC) began the development of a series of standards (19770) to assist organizations in building a SAM program. Each standard in the series was later revised to compliment the growing need for inventory management of software products within an organization. ISO/IEC has, for a long time, been a provider of IT standards that provide a basis from which a defined process can be established.

By establishing a standardized framework for SAM, a basis of effective process management across the industry is established, clarifying the requisite policies and procedures for defining all the process components. Logically, the framework instills into the processes precise management control that is necessary to ensure that all identifiable considerations are made to the software throughout its useful life. Each standard in the 19770 series provides contribution to establish a baseline of software within the organization.

The three standards of the 19770 series pertinent to SAM include:

ISO/IEC 19770-1:2012: "Information technology—software asset management—Part 1: Processes and Tiered Assessment of Conformance" establishes a structured baseline for organizations to implement an integrated set of processes for SAM, which are divided into tiers to allow for incremental implementation, assessment, and recognition. The goal of the standard is to help an organization achieve good practice and manage their software more efficiently so that management, supply chain partners, and other software stakeholders can be confident of the quality and adequacy of the implemented SAM processes. The standard is flexible to the extent that it can be applied to all software and related assets, regardless of their functional purposes. For example, it can be applied to executable application and system software, as well as non-executable software such as fonts, graphics, audio and video recordings. With the growth of cloud computing, the standard has also proven useful in the management of virtualized software applications and other software-as-a-service (SaaS) applications.

ISO/IEC 19770-2:2015: "Information technology—Software Asset Management—Part 2: Software Identification Tag" provides the organization the specifications needed for tagging software to enhance the capabilities needed for identification and management of software assets meeting cyber resilience requirements. Software identification tags, also known as Software ID tags (SWID), help establish the version control of installed software. The tags provide identifying information not only for installed software but also provide a mechanism for the verification of software licensure. It is not uncommon for software vendors to use ISO/IEC 19770-2 as a way to easily, quickly, and accurately identify their software, thus helping staff responsible for SAM process compliance manage their install base more effectively.

ISO/IEC 19770-3:2016: "Information technology—IT asset management—Part 3: Entitlement Schema" provides a set of terms and definitions that can be used when discussing software entitlements. The standard also provides specifications using the extensible markup language (XML) for a defined format of transport, providing digital encapsulation of software entitlement tags (which are files that help identify the software licensing rights). One of the advantages of having a standard for entitlement structure is that it encourages the consistency by industry of names and details of the different types of entitlements. A common definition is critical to standardization and shared understanding.

Identify Facility Assets

Organizational critical services rely on the people, information, software, and technology to support customer and supply chain requirements. Each of these assets requires a physical facility in which to execute the service. The facilities involved in supporting this effort may be shared across multiple services, and ownership may be distributed across many organizations.

As the facility assets are identified, the organization should be conscientious of how that facility affects critical services. Without that consideration, the organization would have difficulty in prioritizing the facilities' effects on those critical services, prior to those housing less significant resources and services. This prioritization should be periodically reviewed and updated on a schedule determined by the organization. Regardless of where the facility stands within the prioritization structure, however during this identification activity all facilities should be recognized.

As was the case with the other asset types, it is essential to identify the primary owner and custodian for the facility assets. The responsibility for a facility asset may not fall under a traditional service area and may be considered an ancillary service to the organization, possibly stretching beyond the organization's boundaries. Another key factor is that when designing organizational services, the key stakeholders may not have considered the importance of the infrastructure or facilities. The organization should ensure that its resilience requirements will be met by any current and future facilities. This point clearly delineates the need for organizational policy that defines the criteria for continuous evaluation of facilities as well as the other three asset types, in terms of their necessity for consideration in the organization's underlying cyber resilience strategies.

Document the Assets

Documentation has been a vital part of organizational success for some time. Clear and decisive specification and guidelines provide an organization the resources necessary to perform critical functions that directly carry out the defined mission and objectives, in addition to the ability to disseminate communication to vital areas of the organization. The importance of documentation is taught very early on in educational institutions. However, the reality of the pace that organizations take in achieving their business objectives has, over the years, pushed the need for documentation much lower in priority. The requirement for substantial documentation is still a component of organizational success, and even more so in managing assets in evaluation for cyber resilience.

In correlation with organizational asset identification and classification of each asset type, it is not uncommon for the documentation discussed in this section to be created and disseminated. The structure of the documentation will vary based on each organization's resilience strategy. Regardless of the structure, however this activity must include the following tasks:

1. Create an asset inventory
2. Document the relationships between identified assets and critical services
3. Analyze dependencies between assets supporting multiple services
4. Continuous update of the asset inventory

Create an Asset Inventory

As assets are identified and mapped to critical services it is important that the organization be able to communicate identification findings to all asset custodians and owners as a means for tracking the satisfaction of resilience requirements. This communication is best provided through an asset inventory. The asset inventory is a document that formally recognizes each asset, how it has been categorized, the services dependent on the asset, and other criteria that serve as a resource for establishing a logical connection between each asset, service, and the organization's resilience strategy.

The means by which the asset inventory is communicated varies by organization. Small to medium size organizations find it beneficial to include records of each asset in electronic format, and it is included within the cyber resilience policy and guideline materials. Large organizations use large database inventory systems to track assets and related services. Queries against the database can then be performed by custodians and owners to examine existing asset/service relationships or perform subsequent updates. Regardless of the recording method used, the inventory typically contains an asset profile (for each identified asset) that includes the following information:

■ asset type (people, information, technology, or facilities)
■ categorization of asset by sensitivity (generally for information assets only)
■ asset location (typically where the custodian is managing the asset)
■ asset owners and custodians (especially if assets are external to the organization)

- format or form of the asset (particularly for information assets that might exist on paper or electronically)
- location of asset backups or duplicates (particularly for information assets)
- services that are dependent on the asset
- value of the asset, either qualitative or quantitative
- asset protection and sustainment requirements

It should be noted that this list of asset profile criteria is not exhaustive. The organization must take the necessary measures to accurately define the right level information based on their own resilience proprieties to document for their own asset base.

Document the Relationships between Assets and Critical Services

Once the inventory with asset profiles has been created, the organization should document what they identified of logical relationships between assets and critical services the asset supports. By including relationship criteria, the organization is able to see not only one-to-one asset/service relationships but one-to many asset/ service relationships will also be highlighted.

In June of 2011, NIST published one method for documenting the assets involved in supporting critical services within their *"Interagency Report 7693, Specification for Asset Identification 1.1."* While the specification was originally developed to provide direction for asset identification within the security automation space, it has since become a valuable guideline for general asset management processes with cyber resilience implications. The specification "describes a framework for how asset management processes and other specifications may identify assets using some set of information known or generated about the asset. It describes the data model and representation of asset identification information and it provides requirements for consuming and producing identification information Requirements for usage of asset information and requirements for how the information" (Wunder et al., 2011).

The Asset Identification specification includes a standardized data model (similar to ISO/IEC 19770-3 for SAM), using XML elements, methods for identifying assets, and guidelines on how to use asset identification. The specification uses several industry-standard mechanisms for representing information consistently. It uses the extensible Address Language (xAL) created by the Organization for the Advancement of Structured Information Standards (OASIS), an XML standard format for representing international address information, in combination with the extensible Name Language (xNL) by OASIS, which is an XML standard format for representing the names of people and organizations. One significant feature of the specification is that it gives examples of how an organization can use identifiers within a database to identify all asset types and relationships between the individual assets and the services they support. For example, a literal identifier can be used to track media access control (MAC) addresses or the real name of a people asset. The specification asserts that a synthetic identifier is then assigned by the organization and can be used to identify not only information but also people, technology, and facility assets. It can also be used as an alternative, in the SAM process, to ISO/IEC 19770. Finally, the specification discusses relationship identifiers so that an asset may be identified based on its relationships to other assets and services.

Analyze Dependencies between Assets Supporting Multiple Services

Generally, critical services do not operate in isolation. Often, service functionality affects the entire organization. Likewise, the assets and resources supported by those services are shared across the organization and are not necessarily dependent on or supportive of just one service area. It is for that reason that it becomes important for the organization to identify these interdependencies in order to ensure that they are addressed during the development of resilience requirements and in the development of strategies aimed at the continuous protection of assets and their related services. Documenting the assets and relationships within the asset profiles gives the organization a baseline from which they can consider whether or not resilience requirements have been implemented at an appropriate level for each individual asset.

When the organization discovers that dependencies exist within a shared environment for an asset, the effects of those dependencies must be analyzed with regard to how well resilience requirements will be satisfied at the service level. The critical service inventory, which is developed as the organization matches services to CSFs, provides an excellent resource that the organization can use to analyze and reveal dependencies between assets and services. It also allows assets to be grouped in multiple ways, such as by service domains, service types, and service components. Analysis of these dependencies and the potential overlap between assets can aid the organization in identifying assets meeting cyber resilience requirements. For example, if resilience requirements are set for a server and that server is utilized by more than one service, the requirements for protecting that server must be sufficient to meet the needs of all services that use it. By identifying these potential conflicts early, an organization can accordingly modify resilience requirements before those assets become an unintentional vulnerability that affects the operational resilience of the affected services.

Update the Asset Inventory

To support its services, an organization should have implemented a defined CM process that can be repeatedly improved to provide more comprehensive asset management. It is through the CM process that the asset inventory should be updated. At its core, CM is a formal process for the rational management of changes to IT products. However, it can often be the foundation for broader service management and facilitate management of the life cycle of those services and their related assets. The overall purpose of the process is to ensure that the status of every meaningful asset is fully documented and known at all times. The goal of the process is to establish and maintain the integrity of every individual asset component by placing them under formal decision making and oversight control by the relevant manager.

By controlling changes in a rational manner, the organization ensures the integrity and correctness of its assets. CM offers two primary advantages to the organization: It maintains the integrity of all configuration items and it allows the appropriate manager to authorize changes to a baseline of those items.

Because CM establishes tangible control over assets and services those assets support, it becomes a vital function of the overall asset management process. The management control applies to all assets throughout their useful life, so CM must be planned and deployed at the organizational level rather than the service level. Because of the differences in focus and requirements between an organization's overall CM and that of an individual service (such as the implications of CM to system and software development projects within the IT function), a distinction must be made between the two.

The goal of service-based CM is to ensure that a customized control strategy is defined and implemented for each critical service of the organization. All service-level assets that will be kept under CM have been identified and formulated into explicit management baselines. Then, a process must be established to ensure that any changes to those baseline components are controlled through proper authorization. CM allows the organization's top-level managers and policy makers to have direct influence over asset changes by involving them in concrete decision making as the structures for performing the services from which those assets support evolve. Also, CM serves as the basis to measure quality by confirming the integrity of asset changes and ensuring that they are not incorporated into the services until they are verified as correct. The latter requirement ensures traceability of changes and enhances problem tracking.

In sum, CM can often be the foundation for broader service management and facilitate management of the life cycle of those services and their related assets. It is important to track and document any changes or updates to an asset throughout its life cycle, keeping in mind potential impacts to the broader service the asset supports.

Manage the Assets

As changes occur in the relationship to identified and classified assets, consideration must be made to subsequent changes required in resilience requirements and the processes that organization employs to ensure that those assets are adequately protected. Likewise, the management of changes that occur in the operational environment (such as the updating of asset inventory and service records in correlation with their requirements) is a critical component of managing and controlling cyber resilience. The organization must consistently monitor all services that satisfy critical objectives for changes that affect the assets being used, identify new assets, or initiate disposal processes for assets that are no longer needed or those that provide a significantly less amount of value to the organization. The main objective, here, is to ensure that the organization's scope for cyber resilience management remains known and controllable.

In order for the necessary level of control to be established, CM becomes the umbrella process from which the assets are managed.

As we iterated in the previous section, CM refers to the understanding and maintenance of information about the status of an organizational asset and the services it supports. CM provides two primary advantages: It maintains the integrity of configurations and allows changes to be evaluated and made rationally. CM also gives the organization's top managers and policy makers direct input into the evolution of the asset base. It does this by ensuring that managers are involved in decisions about the form of the controlled asset. CM provides the basis to measure organizational quality of service delivery, improve the asset life cycle, make testing of each asset against capabilities to support a service and overall organizational Quality Assurance easier, provide traceability of related assets and services, and dramatically ease problems with change management and problem tracking.

Because it establishes the "day one" baseline, the cornerstone of CM is the configuration identification scheme for services and assets. That scheme was described in earlier sections of the chapter in which you were introduced to the tasks of establishing and updating a service catalog and asset inventory. All services and assets are given a unique identifying label and then arrayed, based on their interrelationships and dependencies. The end result of this activity is a baseline which represents the basic configuration of the organization. As our discussions have progressed, we have analogized cyber resilience as locking vital assets in a safe; what is getting "locked up" so-to-speak are these configuration baselines.

Once established, the identification scheme is maintained throughout the asset life cycle and the continuous delivery of each critical service. The organization must explicitly define the management level authorized to approve changes to each baseline. Authorization is always given at the highest practical level. As asset/service dependencies evolve, increasing levels of authority will probably be required to authorize a change. Changes at any level within the organization must be maintained at all levels. Therefore, it becomes necessary for the organization to create a Configuration Control Board (CCB) that operates at defined levels of authorization. CCBs are hierarchical and composed of managers with sufficient authority to direct the change process. At a minimum, an organization has six control boards: one composed of top-level policy makers and one for each of the major system components

Figure 2.9   Tasks associated with managing assets.

(a human resource CCB, a facilities CCB, a software CCB, a technology CCB, and information CCB). The members of these boards have the proper level of authority to oversee decisions. The scope of the board's oversight must be formally and explicitly defined, usually in the general CM plan of the organization (Figure 2.9).

Specifically, the tasks associated with managing assets include

1. identifying change criteria;
2. establishing a change schedule;
3. managing changes to assets and inventories;
4. improving the process.

Identify Change Criteria

As we alluded to in our discussion of CM, organizational assets are dynamic. As assets change, their resilience requirements and protection necessities also change. For the organization to effectively manage its assets, it must use the foundations of CM to actively monitor for changes that significantly alter assets, identify new assets, or request the initiation of the process for disposal of assets no longer needed.

To establish an efficient means for evaluating changes to assets, the organization must develop a set of change criteria that is consistently applied through CM within the organization. The change criteria should cover all the asset types: technology, people, information, software, and facilities. Through the effective use of change criteria, changes within assets can be translated into new baselines that in turn change the resilience requirements. To confirm that change criteria are consistently applied, official documentation (aside from the inventories previously discussed) should also be created in the form of policies and procedures based on previously defined organizational documentation standards.

As mergers continue to be common in many industries, we can use that business practice to illustrate change. Consider an automotive manufacturer which has just acquired a smaller automobile manufacturer. Changes in the organizational structure will occur to ensure that the structure of the acquired manufacturer aligns with the mission of the acquirer. This will require the people in the acquired manufacturer to receive specialized training. Additionally, facilities of both organizations must also be merged, and resilience requirements such as security practices will need to be implemented at those facilities. Having identified change criteria following a set of documented policies and procedures will greatly assist the new organization identify and track changes to its asset inventory and ensure that resilience requirements are captured.

Establish a Change Schedule

Effective planning is the cornerstone in all business management practices. Within each of those plans is generally a schedule that provides the direction an organization needs to accomplish tasks in a timely manner. Thus, management should create and enforce measures for updating assets and reviewing resilience requirements as a means for managing change. To start, the organization should consider linking the review of technology, software, information, and facility assets to the system development life cycle (SDLC). The benefits for linking the SDLC to asset management include

- the ability to asset milestones such as software testing completion or facility upgrades;
- the chance that vulnerabilities can be identified in assets early on;
- the ability to identify potential challenges in the design of assets through the review process of the SDLC;
- the availability of documentation created based on resilience decisions made within each SDLC phase.

Of course, changes and review of resilient requirements of software, information, technology, and facilities goes well beyond IT implications and should also be considered for other critical services throughout the organization. Scheduled changes and reviews to resilience requirements of people assets require a slightly different approach because of the scope of people affecting each service throughout the organization. HR management becomes a factor in the use of common tools like performance reviews that can be used to document the skills of personnel and identify training needs. The organization should determine the frequency at which these reviews should take place.

Manage Changes to Assets and Inventories

The internal and external circumstances under which an organization is expected to operate are continually changing. As a result, the risk environment for services and their supporting assets continues to evolve as well. An organization must become very proficient at recognizing changes in conditions that precipitate considerations for changes in asset resilience requirements. It is through CM that those changes can be effectively managed, documented, and scheduled for implementation.

Using CM as the vehicle for managing change provides the means by which the recording, retrieval, and maintenance of the current asset configuration and all preceding configurations are kept under rigorous management control to assure correctness, timeliness, integrity, and security. The status of every asset is documented to become part of an explicit and coherent baseline. This baseline represents the status of the asset at a fixed point in time or circumstance.

Once the asset baseline has been established, the rationale for changing it and any associated authorizations are described in the configuration record and maintained throughout the asset life cycle. All changes to configuration baselines are recorded, evaluated, and approved as they are requested. Once a change is made, the change is verified and incorporated into a new configuration baseline and the old baseline is archived. Audits should be performed as needed to verify that baselines conform to their resilience requirements.

Improve the Process

Without a shred of doubt, asset identification and classification for cyber resilience is an ongoing process for the organization. As new CSFs are determined and the organization's services evolve as do supporting processes, so must the assets that support those services. The organization should continually assess these changes to properly manage and improve on decisions related to cyber resilience. However, there is no "silver bullet" into how improvement takes place. Each organization is unique with regard to how the critical services are defined and resilience requirements prioritized, to the extent that improvement needs and steps to accomplish them will also be unique. Nevertheless, a plan must be developed that provides details as to how changes to the process will be designed, implemented, and monitored.

While developing their plan for improvement, they should carefully consider the details we have provided in this chapter in combination with all other facets of cyber resilience management. Lessons learned from the deployment or management of assets in different parts of the organization may provide information that will enable the organization to manage those assets uniformly. Other cyber resilience processes to consider within improvement planning include:

- Controls Management—As the priorities and objectives of the organization change, a decision may be made to use different security controls to add or dispose of technology assets more efficiently. The controls necessary to achieve such efficiencies should be fed into the asset management process so that assets that support these controls can be implemented.

- Incident Management—The assets used during an incident should be discussed during the post-incident reporting. Recommendations to improve the management of the assets should be made.
- Risk Management—The organization's normal risk assessments will generally reveal new risks, which might be mitigated by considering them within the asset identification and classification process.
- Business Continuity—As disaster recovery and business continuity plans are developed and exercised, failures should be documented and recommendations for new assets should be considered within the asset identification and classification process.

Chapter Summary

We are living in a period of unprecedented technological change. Building resilience to these changes is becoming increasingly imperative. Building resilient organizations that can provide solutions and adapt to these new challenges will be a major task in the coming years. To become more resilient in this day and age of continued digital disruption increasingly means understanding the full scope of cyber governance responsibilities. This means starting with a top-down approach in managing risk. There is no point in identifying what will make the organization more resilient until there is complete understanding of the core mission and objectives. From there, a decisive understanding of what services are critical to that achievement of the mission and objectives can be established.

Critical services require a continuous supply of critical assets that provide the means by which continuous service delivery can be sustained. Organizations need to identify their most critical assets and have alignment with the critical services, down to the individuals who are responsible for protecting them. Organizations must assess what people, technology, software, facilities, and information are critical, where they are stored, how they flow across the organization, and who really needs access to them. Furthermore, organizations must recognize the impact on the business should these critical assets be compromised and be prepared to respond to limit the impact to the organization while restoring normal business operations.

However, simply knowing what assets support critical services is not adequate. CM is the vehicle by which assets that meet cyber resilience requirements are documented and managed. As business priorities change, so to do the services that satisfy those priorities. Changes within all levels of the organization can directly or indirectly affect changes to service supportive assets. Organizations that have a defined CM process are better equipped to manage those changes and accurately reconfigure asset baselines.

Keywords

Asset custodian: A person or organizational service area, internal or external to the organization, that is responsible for satisfying the resilience requirements of an identified critical asset

Asset owner: A person or organizational service area, internal or external to the organization, that is responsibility for the viability, productivity, and resilience of an asset

Asset management: A management paradigm and a body of management practices that is applied to the entire portfolio of assets and aims to minimize the total cost of acquiring, operating, maintaining, and renewing the assets while continuously delivering the service levels customers desire and regulators require at an acceptable level of business risk to the organization

Configuration management: A formal process to ensure the continuing status of a logically related array of components

Critical infrastructure: The bodies of systems, networks, and assets that are so essential that their continued operation is required to ensure the resilience of an organization

High-value services: Those services within an organization, whose continuous operation is critical to successful accomplishment of the organization's mission

Provisional impact value: A metric for determining the level by which an information asset is subjected to confidentiality, integrity, or availability requirements

Software entitlement: Enabling an organization to define the people or machines to which a specific, purchased software license is assigned

References

Microsoft. (n.d.), Software asset management. Accessed Sep 23, 2017. https://www.microsoft.com/en-mt/sam/.

National Institute of Standards and Technology. (2014), *Framework for Improving Critical Infrastructure Cybersecurity*, NIST, Gaithersburg, February 12.

Wunder, J., Halbardier, A., and Waltermire, D. (2011), *NIST IR7693: Specification for Asset Identification 1.1. Specification*, National Institute of Standards and Technology, Gaithersburg.

Chapter 3

Establishing the Risk Status of the Corporate Infrastructure

At the end of this chapter, the reader will understand

1. what threat modeling is;
2. security requirements;
3. approaches to threat modeling;
4. types of threat actors;
5. diagrams using the Unified Modeling Language (UML);
6. threat modeling methodologies;
7. the components of a sample threat model.

Introduction

Threat modeling is something that is widely engaged in but often taken for granted. In other words, many people threat model yet they may not realize that is what they're actually doing. For example, say you're planning a road trip by car that will encompass roughly 600 miles one way. Before you hit the road, you build an implicit threat model where you consider if your tires have enough traction and are safe for dealing with the possibility (threat) of rain or snow; you may consider the condition of the engine, fluid levels, and safety features to ensure they are all in good working order (threat of possible breakdown); if you happen to speed, you may consider getting a radar detector to alert you to the possible threat of a police officer pulling you over or you may have a ready speech for the officer to attempt to get out of the threat of an expensive ticket. Threat modeling is the process of analyzing what might go wrong and to consider all the possibilities of what could happen for any given situation.

Threat modeling involves a variety of skills and knowledge about the organization such as

- an understanding of the organization's security requirements;
- knowledge of the organization's mission critical assets and the owners of the data such as department managers or application teams;
- knowledge of databases, data flow diagrams (DFDs), and where the data is stored;
- knowledge of network architecture, infrastructure components, entry points, and technologies in place to continuously monitor the network traffic;
- knowledge of what security controls have been implemented and, ideally, knowledge of security controls that have been violated or are routinely ignored;
- understanding of risk management to include risk identification, risk estimation, evaluation of risk levels and acceptance criteria, risk treatment options, and residual risk assessment.

This chapter will address the importance of threat modeling and discuss strategies focusing on assets, attackers, and software to find threats and provide defensive tactics to address threats. A comprehensive sample threat model will also be provided so that the reader can understand all the facets of threat modeling.

Security Requirements

Organizations are struggling with how to address cybersecurity issues and finding knowledgeable and capable talent to do so. And throwing money at the problem without understanding the unique security requirements of the organization will not help. It is critical to start at the beginning and understand the business, specifically the business requirements; essential processes; mission critical systems and data; activities that must be performed to meet business objectives and goals; legal and governmental regulations; and customer needs/expectations. This information is then used to determine functional security requirements such as:

- *Access control*—It involves determining who (employees, contractors, suppliers, etc.) and what (systems, interfaces) requires access either *physically* (to information technology (IT) assets, rooms, buildings, or campuses) or *logically* (to system files, networks, data) to organizational resources. Access controls perform the necessary function of identifying and authenticating who or what can view or use resources in the computing environment.

- *Privacy*—The organization may need to support anonymity, pseudonymity, or unlinkability, however serious consideration of non-repudiation must be considered. Non-repudiation involves providing proof of the integrity and origin of data to ensure that a communication or party to a contract cannot deny the authenticity of sending a message they originated or a document they signed. As you can discern, non-repudiation and privacy can seriously conflict with one another.
- *User data protection*—Policies on access control and data information flow rules must be in place to support the transmission of user data, data at rest, off-line storage, and residual information protection such as the ability to recover "deleted" data.
- *Security audits*—It includes the identification of what the organization wants to make auditable, what activities get recorded, stored, and analyzed.
- *Trusted path/channels*—It involves establishing a way in which users can ensure they are connecting to the real site, as hackers use forged login screens that mock corporate web pages.
- *Data integrity*—It involves the assurance of the accuracy and consistency of data over its entire life cycle. This can be accomplished during the database design phase when entity, referential, and domain integrity constraints are implemented.
- *Availability*—It involves ensuring that software applications, tools, systems, and information are available when and where it is rightly needed.

There is a significant interdependency between functional security requirements, threats, and how the organization chooses to mitigate those threats. As you model threats, decisions will have to be made as to whether the cost to address the threat exceeds the benefits. Some threats may not be worth addressing and understanding the organization's business requirements and strategic plans will help guide those decisions. It is important that whatever decisions are made to address threats, they are fully documented and communicated.

Approaches to Threat Modeling

There are two common approaches to structuring a threat model: (1) focusing on critical assets and (2) focusing on software applications, both in stages of development and software applications that are in production. A third, but less common approach, focuses on potential human-centered attackers and requires research and experience to develop the profiles of attackers. We will discuss this only briefly as this is a much less effective approach and not commonly used.

Critical Asset Approach

Critical assets are those that if breached would have a significant negative impact on the mission and/or operation of the organization. Examples of critical assets are databases, systems, software applications, business processes, and will vary widely across sectors. For financial-services organizations, transactions and their mergers and acquisition database(s) will require more protection than their marketing materials. For hospital systems, the most sensitive asset is generally their patient information. An aerospace manufacturer needs to protect their intellectual property as their critical asset. Hence, an understanding of the business and the data flow to include all systems and networks that transport that data would be in the purview of a threat modeler.

Determining which assets are critical is the first step in understanding what needs to be protected and how to mitigate any potential threat. Assets that impact confidentiality, integrity, availability, the mission of the organization, and business functions all need to be identified and assessed to consider how an attacker could threaten each asset. Many organizations track hardware and software using asset tracking applications, which is a good way to start identifying what is in use or production. Network traffic monitoring tools can be very helpful in understanding the most used network and system components, and once this information is obtained, additional information can be acquired such as identifying authorized users, contractors, strategic business partners, power users, and high-risk users who most often interact with the assets.

The next step is to diagram the mission critical assets and systems using the UML standards to show the assets and interconnections with the network to understand how those assets could be attacked. More information on UML diagramming will be presented later in the chapter.

Software Application Approach

Software-centric models focus on software applications in development as well as systems to be implemented in a production environment. Depending on the development methodology being employed, project documentation of the systems and applications will differ. For example, for larger complex projects, it is more common to use a methodology where the design of the database, business rules, logic, etc. are well documented. For organizations using agile methodology or extreme programming, the software and systems may not be documented. UML standards, such as use cases, DFDs, logic tables and/or trees, sequence diagrams, activity diagrams, class diagrams, and entity relationship diagrams, can be useful in ensuring the project team understand the potential threats and who will address any security issues. Too often, developers focus on making the software function properly, and security is an afterthought.

Over time, projects increase in complexity and it is helpful for project teams to use a whiteboard to diagram how the variety of production software fit together to expose any potential security issues. Process mapping can be incredibly enlightening when multiple project teams have different understandings of how their projects fit into the overall big picture of the corporate environment. For those who may not have performed whiteboarding exercises, know that discussions can be rather lively at times, but it is very worth it in the end.

It is also effective to combine both the software development teams with the network teams to understand the infrastructural components such as platforms, routers, switches, and how both the software and network infrastructure fit into the established trust boundaries. DFDs are useful for diagraming which IT resources with different privileges interact and which resources change their levels of "trust" to attain increased privilege levels. The distinct boundaries within which a system trusts all subsystems and data should be understood and protected as threats tend to cluster around trust boundaries. Typically, software developers, network architects and administrators have significantly different skill sets and bringing them together can enhance project designs.

Attacker Approach

The attacker approach focuses on human threat agents and starts with the development of attacker lists with the idea of making threats appear more real. An attacker list is challenging to create in that starting from a list of attacker archetypes, serious thought as to what resources and capabilities each archetype may possess, results in a more detailed attacker persona. Persona development is an intensive data and research driven process. Personas provide a way of thinking about how users (threat agents) behave, how they think, what they wish to accomplish, and why (Cooper et al., 2014). They are based on the behavior and motivations of real people observed and gathered from many ethnographic interviews. Researchers Reimann, Goodwin, and Halley (Cooper et al., 2007, p. 97), developed a seven-step process for creating descriptive models of users called *personas*:

1. Identify behavioral variables
2. Map interview subjects to behavioral variables
3. Identify significant behavior patterns
4. Synthesize characteristics and relevant goals
5. Check for completeness and redundancy
6. Expand descriptions of attributes and behaviors
7. Designate persona types

Persona modeling was developed to satisfy the users of products and aid in the product development design process, however this same process can be used to develop models of threat agents.

Another persona classification model, created by David Aucsmith (Shostack, 2014; Aseef et al., 2005), is a multidimension schema based on two key elements: motivation and skill level. In this classification model, Aucsmith identifies five threat agents, shown in Figure 3.1: vandal, author, trespasser, thief, and spy, and depending on their motivation and skill levels, patterns of behavior can be identified that describe emerging threat personas who engage in threat scenarios. The following diagram illustrates the matrix of motivation (ranging from curiosity,

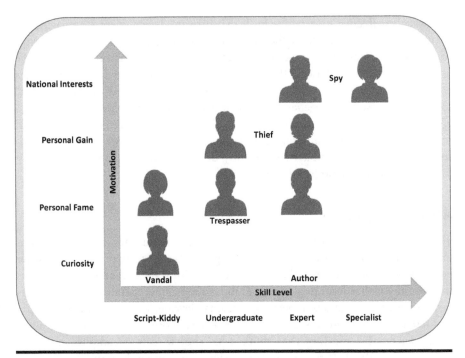

Figure 3.1 Aucsmith's threat persona classification model.

personal fame, personal gain to national interests) coupled with skill levels (ranging from script kiddy, undergraduate, expert to specialist) and the threat agent persona possibilities.

Types of Threat Actors

Verizon (2018) publishes a *Data Breach Intelligence Report* each year and from their data three types of threat actors are identified, but differ depending on the industry (health care, retail, manufacturing, public, etc.):

1. *Insider threats*—28% of attacks involved insiders and included both malicious employees who were looking to line their pockets and negligent employees who made errors such as failing to shred confidential information, sending an email to the wrong person, or misconfiguring web servers (Verizon, 2018).
2. *Cybercriminals/organized criminal groups*—These include individuals who are looking to do much damage and make more money by encrypting file servers and databases (as opposed to single user devices) by creating and deploying ransomware; payment card skimmers installed on ATMs; Point of Sale (POS) intrusions; and social engineering/phishing expeditions. Members of

this group were behind half of all 2,216 confirmed data breaches analyzed in the report (Verizon, 2018).

3. *Nation-state/state-affiliated*—12% of attacks were committed by this member group and included cyberespionage and attacks involving phishing, installations and the use of backdoors or C2 channels to target state secrets and personal data of citizens and employees (Verizon, 2018).

Intel Threat Agent Library

In 2007, the IT Threat Assessment Group at Intel Corporation developed a standardized threat agent library (TAL) to provide a consistent reference describing the human agents that post threats to IT systems and information assets (Casey, 2007). Figure 3.2 lists the variety of threat agents identified by the Intel IT Threat Assessment Group, their common tactics, and a description of each agent label.

Two additional threat actor types to add are *hactivists*, who are not necessarily motivated by money and generally have a political agenda. Their goal is to cause damage to organizations or individuals whom they are opposed to in order to gain awareness of their issues. *Opportunistic* threat actors look to profit from finding flaws and exposing exploits in network systems and devices. They range from amateur criminals to professional hackers who may be driven by the desire for notoriety. Some genuinely want to help organizations find security vulnerabilities, however many are looking for bragging rights and personal gain.

After the threat agent list has been developed, more detailed information can be researched for the creation of attacker personas. Figure 3.3 provides a sample persona profile of an individual who may pose a threat to an organization.

Diagrams Using the Unified Modeling Language

Diagramming your network and software applications can be an invaluable way to gain a clear understanding of the design and components that have been deployed, trust boundaries, and security measures that may be in place, especially for complex projects. Although there are many ways to diagram, we prefer to use the UML standard. There are two broad categories of diagrams within this standard: *behavioral* and *structural*. Behavioral UML diagrams include use/misuse case diagrams and scenarios, activity and sequence diagrams, communication and state diagrams. Structural diagrams include class, object and entity relationship diagrams, DFDs, process maps using swimlanes, and component diagrams.

The goal of all these diagrams is to accurately depict how each area works as well as how the overall system works together with the important factor that this gives everyone a clear understanding of the same thing. Large and complex projects are often divided into smaller projects, which may have different teams working

	Agent Label	Insider	Common Tactics/Actions	Description
Hostile	Anarchist		Violence, property destruction, physical business disruption	Someone who rejects all forms of structure, private or public, and acts with few constraints
	Civil Activist		Electronic or physical business disruption; theft of business data	Highly motivated but non-violent supporter of cause
	Competitor		Theft of IP or business data	Business adversary who competes for revenues or resources (acquisitions, etc.
	Corrupt Government Official		Organizational or physical business disruption	Person who inappropriately uses his or her position within the government to acquire company resources
	Cyber Vandal		Cyber Vandal Network/computing disruption, web hijacking, malware	Derives thrills from intrusion or destruction of property, without strong agenda
	Data Miner		Theft of IP, PII, or business data Professional data gatherer external to the company (includes cyber methods)	Professional data gatherer external to the company (includes cyber methods)
	Employee, Disgruntled	X	Abuse of privileges for sabotage, cyber or physical	Current or former employee with intent to harm the company
	Government Spy	X	Theft of IP or business data	State-sponsored spy as a trusted insider, supporting idealistic goals
	Government Cyberwarrior		Organizational, infrastructural, and physical business disruption, through network/computing disruption, web hijacking, malware	State-sponsored attacker with significant resources to affect major disruption on national scale
	Internal Spy	X	Theft of IP, PII, or business data	Professional data gatherer as a trusted insider, generally with a simple profit motive
	Irrational Individual		Personal violence resulting in physical business disruption	Someone with illogical purpose and irrational behavior
	Legal Adversary		Organizational business disruption, access to IP or business data	Adversary in legal proceedings against the company, warranted or not
	Mobster		Theft of IP, PII, or business data; violence	Manager of organized crime organization with significant resources
	Radical Activist		Property destruction, physical business disruption	Highly motivated, potentially destructive supporter of cause
	Sensationalist		Public announcements for PR crises, theft of business data	Attention-grabber who may employ any method for notoriety; looking for "15 minutes of fame"
	Terrorist		Violence, property destruction, physical business disruption	Person who relies on the use of violence to support personal socio-political agenda
	Thief	X	Theft of hardware goods or IP, PII, or business data	Opportunistic individual with simple profit motive
	Vendor	X	Theft of IP or business data	Business partner who seeks inside information for "financial advantage over competitors
Non-Hostile	Employee, Reckless	X	Benign shortcuts and misuse of authorizations, "pushed wrong button"	Current employee who knowingly and deliberately circumvents safeguards for expediency, but intends no harm or serious consequences
	Employee, Untrained	X	Poor process, unforeseen mistakes, "pushed wrong button"	Current employee with harmless intent but unknowingly misuses system or safeguards
	Information Partner	X	Poor internal protection of company proprietary materials	Someone with whom the company has voluntarily shared sensitive data

Figure 3.2 Intel Corporation summary of threat agent information.

David Defacer

Core Profile	
Motivation	• Personal fame, political purposes motivated by ideology
Background, Skill & Education	• Moderate programming and network experience • Relatively sophisticated and skilled at being able to modify original exploit and add code • Experience in configuring standard networking components • Is the 'computer expert' for family and friends • Frequently plays video games • Excelled in information technology courses in HS • Currently enrolled in an undergraduate IT program
Job & IT Experience	• Worked part-time as a computer field support technician and was promoted to junior network administrator for small company
Previous Accomplishments	• Planted a Sub7 Trojan horse program on open computers at his former workplace to protest their views on social justice causes • Destroyed data from his former employer's website when the Trojan horse he implanted cause the web server to fault • Successfully launched a DDoS attack against several websites that supported social justice causes he disagreed with • Wrote a few successful exploits against IIS and Apache
Span of Influence	• Limited to his group who share ideological views
Scope of Attacks	• Former employer and other companies headquartered in California

Figure 3.3 Sample attacker persona profile.

on them, and it is not uncommon for one team to have limited knowledge of what another team is doing. Diagraming all the components helps get everyone involved on the same page, and using an industry standard limits the amount of reexplaining as people cycle on and off the project teams. These diagrams can often "say" things that would be difficult to explain in a narrative and work for both technical and nontechnical individuals.

Although there are many diagrams to choose from, each one serves a different purpose. Every organization may favor a particular set of diagrams, but one of the most used for threat modeling, and one of the best, is the DFD. DFDs can be created for both software applications and architectured systems as problems tend to follow the data flow. DFDs map out the flow of information for any process or system and can range from simple high level to in-depth that dig progressively deeper into how the data is handled. DFDs use defined symbols such as rectangles with sharp corners to depict *entities*; rectangles with the right side of the rectangle left open or two parallel lines with a label to depict *data storage units*; arrows to depict *flows* between processes and data stores; rounded squares to depict *processes*; and circles to depict *systems*. Figure 3.4 shows a sample DFD.

There are several levels of DFDs with the highest level called *level 0* or a *context level* diagrams, which depict a very broad view with limited detail. A classic level 0 DFD should show only a single process and the connections that a process has with external entities. We say classic in the sense of UML guidelines, but we've seen DFDs drawn on paper napkins with systems thrown in for good measure. DFDs can be depicted focusing on the *logical* elements such as the business and business activities or a *physical* diagram that focuses on how the system is implemented. Some organizations produce a mixture of the two. It all depends on what the norms are in your organization. *Level 1* DFDs depict more detail to include other relevant processes that are involved in the flow of data. Additional data flows, data stores, and systems can be depicted with each process and data store numbered. It is best to limit the amount of processes on each page to reduce potential confusion. By numbering each process, another diagram, *level 2*, can be created that depicts the subprocesses. With a numbering scheme, it will be easier to link all the diagrams together as a whole to understand where the trust boundaries are. As you diagram out your network, system, or software application, it is important to identify where the levels of privilege change as threats tend to cluster around trust boundaries. The trust boundaries define the attack surface between the entities, and clarity about what boundaries exist will allow you to build good threat models. Good threat models will then allow the team to work on how those boundaries are to be best protected.

Process maps with swimlanes are another useful diagram in helping get everyone on the same page. In addition to showing each step of the way, decisions that are made are usually *yes* or *no,* and depending on the answer of the decision, an arrow

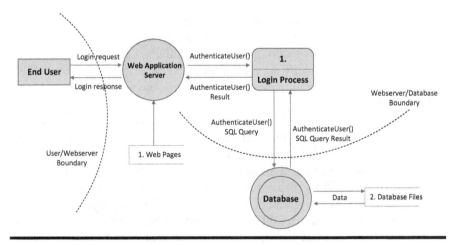

Figure 3.4   Sample DFD.

depicts the flow to the next step. Swimlanes help separate the different functions or entities involved and each lane is labeled with the function, participant, place, or thing. Figure 3.5 shows a classic process map with swimlanes.

Use case diagrams are organized collections of scenarios that are derived from the business rules and actions taken by the various users of a system. They focus on what the system can do or its purpose within the organization. Use cases map out the actors, business activities or processes, flow of events, and post-conditions or alternative paths, and focus on the functionality of the components of a software application or system. They can be helpful ways to graphically illustrate the business requirements and who will be using the various functionalities of the software or system. In threat modeling, it is helpful to take a use case diagram and extend them to add the identified threats in which an attacker may engage. The addition of threats to a use case turns it into a misuse case diagram. An example of a misuse case diagram can be seen in Figure 3.6.

Diagramming complex systems can be daunting in order to get a full understanding of where the trust boundaries exist and everything involved. One rule of thumb is to balance the highly detailed areas with consistently numbered context, level 1 and level 2 diagrams.

Here are some suggestions as to what to include in your diagrams, shown in Figure 3.7 (Shostack, 2014; Howard and LeBlanc, 2009):

■ Show the events that drive the system.
■ Show the processes and subprocesses.
■ Determine what responses each process will create and send.

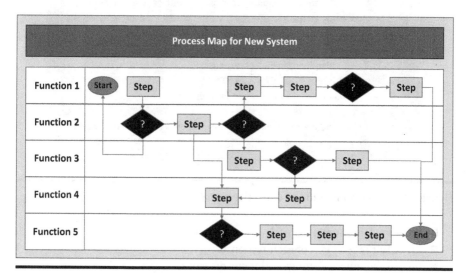

Figure 3.5 Sample process map with swimlanes.

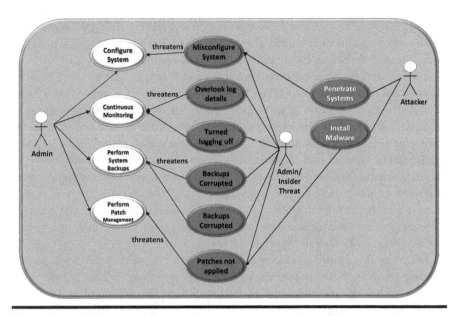

Figure 3.6 Misuse case diagram.

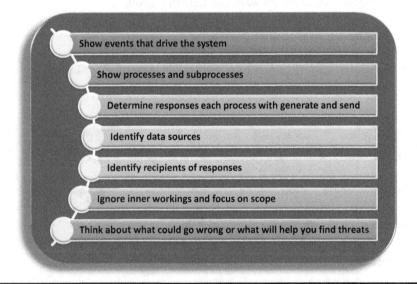

Figure 3.7 What to include in your diagrams?

- Identify data sources for each request and response.
- Identify the recipient of each response.
- Ignore the inner workings and focus on scope.
- Think about what could go wrong or what will help you find threats.

Threat Modeling Methodologies

Once you understand the organization's security requirements and the variety of potential threat agents, select a threat modeling approach that suits your organization; armed with current system diagrams, you can then begin to find threats. There are several models or techniques that can aid in finding risks and provide structure for your threat model. They are STRIDE, DREAD, OWASP's (Open Web Application Security Project) Top Ten Project, attack trees, and attack libraries.

STRIDE

STRIDE is an acronym that stands for Spoofing, Tampering, Repudiation, Information Disclosure, Denial of Service, and Elevation of Privilege as shown in Figure 3.8. Developed by Microsoft, STRIDE is a threat classification model designed to help identify security threats in six categories in order to help answer the question: "what can go wrong in this system we're developing or working on?" (Shostack, 2014). The model can be used to find threats against software or can be

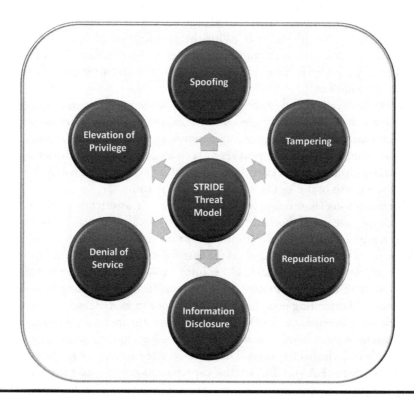

Figure 3.8   STRIDE threat model.

used more broadly to encompass processes, data stores, data flows, as well as potential attacks against trust boundaries.

A *spoofing* attack is basically pretending to be someone or something other than who you actually are and can be a person, machine, file, website, process, or even a role. To spoof a person or user identity, attackers often take over a real email account or create a new email account, LinkedIn profile, Twitter, or Facebook profile using the victim's information. It is imperative for applications to verify the identity of the sending or receiving host as many of the Transmission Control Protocol/Internet Protocol (TCP/IP) protocols do not provide a mechanism for authenticating the source or destination of a message. Firewalls with deep packet inspection functionality can help mitigate a spoofing attack. Spoofing a machine can be accomplished at several layers of the Open Systems Interconnection (OSI) model such as IP address spoofing and domain name system (DNS) spoofing or DNS cache poisoning.

A *tampering* attack involves modifying memory, files, or even data flowing over a network. Memory can be modified if an attacker gets access to code and shares the same privilege level. An application programming interface (API) can be modified after a security check is performed if the API is configured to handle data by reference. If an attacker has write permissions, they can modify files both locally and remotely. They can also modify links or redirects, which are often left out of integrity checks. With the proliferation of open wireless and Bluetooth networks, network tampering has become easier. Attackers often alter web logs and log files in order to hide their activities and attempt to remain on a network unnoticed.

A *repudiation* attack can happen when an application or system does not employ the proper controls to track or log user actions, thus allowing users to deny that they performed specific actions or transactions. To overcome repudiation threats, strong authentication, logging features to keep an audit trail of activities, and use of digital signatures are necessary. It is also helpful to define what will be logged to aid in the analysis of the logs to handle repudiation issues.

Information disclosure is simply allowing people to inadvertently see information in which they are not authorized to see. This can include exploiting a website that reveals sensitive data such as error messages or developer comments and directory indexing attacks, which can happen if the base file is not present, all the files within a requested directory will be listed. Other information disclosures can include finding crypto keys on a disk, in memory, or in emails; obtaining data from logs or temp files; and exploiting poor database permissions to gain access.

A *denial-of-service* attack is intended to cause disruption by flooding a targeted machine or network resource with excessive requests so as to render it unavailable to its legitimate users. Denial-of-service threats can be against a process to absorb the RAM, CPU, or disk capacity or against a data store in which enough requests are generated to shut down the system. A distributed denial-of-service is where multiple compromised systems flood the targeted resource, such as bandwidth or a file server, with traffic again, with the purpose to overload and shut it down.

An *elevation of privilege* attack occurs when an attacker is able to obtain authorization permissions beyond those initially granted. For example, an attacker with an initial privilege set of "read only" permissions is somehow able to elevate the set to "read and write" or even gain "administrator" or "super user" access. Elevation of privileges can be obtained by corrupting a process by sending inputs that software code cannot properly handle. Another way to elevate privileges is through failure to not track authorization in every phase of the application. If the application relies on other programs, datasets, or configuration files, it is important to ensure that permissions are set on each of those dependencies so that they are trustworthy.

DREAD

Used by OpenStack's Security Group, DREAD is a risk assessment scoring model to understand the potential impact and severity of vulnerabilities of their cloud computing deployments (OverStack, 2018). The DREAD model has five categories, with each having scores ranging from zero to ten, where zero indicates no impact and ten is the worst possible outcome. All five categories are summed together and then divided by five to show the overall risk of the vulnerability:

$$Risk = (Damage + Reproducibility + Exploitability \\ + Affected\ Users + Discoverability)/5$$

Damage potential is scored based on if the vulnerability is exploited, how much damage would be caused:

- ■ 0 = Nothing.
- ■ 3 = Individual user data is compromised, affected, or availability is denied.
- ■ 5 = All individual tenant data is compromised, affected, or availability is denied.
- ■ 7 = All tenant data is compromised, affected, or availability is denied.
- ■ 7 = Availability of a specific cloud controller component/service is denied.
- ■ 8 = Availability of all cloud controller components is denied.
- ■ 9 = Underlying cloud management and infrastructure data is compromised or affected.
- ■ 10 = Complete system or data destruction, failure, or compromise.

Reproducibility is scored based on how reliably the vulnerability can be exploited.

- ■ 0 = Very hard or impossible, even for administrators. The vulnerability is unstable and statistically unlikely to be reliably exploited.

- 5 = One or two steps required, tooling/scripting readily available.
- 10 = Unauthenticated users can trivially and reliably exploit using only a web browser.

Exploitability is scored based on how difficult it is to exploit the vulnerability. Given enough time and effort, all vulnerabilities are exploitable and so, all scores should be scored between one and ten.

- 1 = Even with direct knowledge of the vulnerability there is no viable path for exploitation.
- 2 = Advanced techniques required, custom tooling. Only exploitable by authenticated users.
- 5 = Exploit is available/understood, usable with only moderate skill by authenticated users.
- 7 = Exploit is available/understood, usable by non-authenticated users.
- 10 = Trivial - just a web browser.

Affected Users is scored based on the number of users that may be affected.

- 0 = None.
- 5 = Specific to a given project.
- 10 = All users impacted.

Discoverability is scored on the ease with which a threat may be discovered or detected. Some security analysts omit this category and advise that Discoverability should be at its maximum rating so as not to reward security through obscurity as the main method of providing security.

- 0 = Very hard to impossible to detect even given access to source code and privilege access to running systems.
- 5 = Can figure it out by guessing or by monitoring network traces.
- 9 = Details of faults like this are already in the public domain and can be easily discovered using a search engine.
- 10 = The information is visible in the web browser address bar or in a form.

Figure 3.9 provides an example of a vulnerability score using DREAD.

OWASP's Top Ten Project

The OWASP is an open community that provides tools, documents, videos, presentations, and chapters to improve software application and API security. Their most recent document, *OWASP Top 10—2017*, identifies the ten most critical web application security risks and is based on 40+ data submissions from firms that

Potential for: Tampering, Escalation		
Category	**Score**	**Rationale**
Damage	6	Significant disruption
Reproducibility	8	Code path is easily understood, condition exists as standard
Exploitability	2	Very hard to exploit without specific conditions
Affected Users	8	All cloud computing users
Discoverability	10	Discoverability always assumed to be 10
DREAD SCORE: 31/5 = 6.2 – Important, fix as a priority		

Figure 3.9   Sample vulnerability score using DREAD.

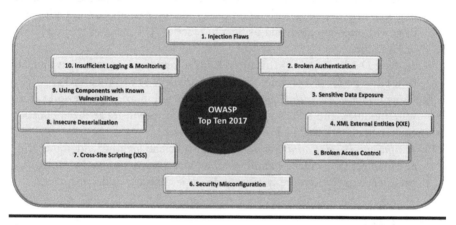

Figure 3.10   OWASP's Top Ten 2017.

specialize in application security and over 500 completed surveys of industry professionals. The data spans "vulnerabilities gathered from hundreds of organizations and over 100,000 real-world applications and APIs" (OWASP, 2017). The top ten application security risks, identified in the *OWASP Top 10—2017* document and shown in Figure 3.10, are briefly defined here:

1. *Injection Flaws*—They occur when untrusted data is sent to an interpreter as part of a command or query allowing an attacker to access data without proper authorization.
2. *Broken Authentication*—It occurs when application functions related to authentication and session management are implemented incorrectly allowing attackers to compromise passwords, keys, session tokens, or assume other users' identities.

3. *Sensitive Data Exposure*—Attackers may steal or modify poorly coded APIs, data at rest, in transit, and data exchanged with web browsers if not properly protected.

4. *External Entities (XXE)*—XML documents may be poorly configured allowing external entities to disclose internal files using the uniform resource identifier (URI) handler, internal file shares, internal port scanning, and remote code execution.

5. *Broken Access Control*—It occurs when restrictions on what authenticated users are allowed to do are often not properly enforced.

6. *Security Misconfiguration*—This is the most commonly seen issue; systems must be kept up-to-date with patches and upgrades. Additional issues such as insecure default configurations, incomplete or ad hoc configurations, open cloud storage, misconfigured HTTP headers, and verbose error messages containing sensitive information are included in this category.

7. *Cross-Site Scripting (XSS)*—It occurs when an application includes untrusted data in a new web page without proper validation or updates an existing web page with user-supplied data using a browser or API that can create HTML or JavaScript.

8. *Insecure Deserialization*—This often leads to remote code execution and can be used to perform replay, injection, and privilege escalation attacks.

9. *Using Components with Known Vulnerabilities*—Libraries, frameworks, software modules, and APIs running with the same privileges as the application can facilitate serious data loss or server takeover.

10. *Insufficient Logging and Monitoring*—Insufficient logging and monitoring and ineffective incident handling allow attackers to maintain undetected persistence, pivot to more systems and tamper, extract, or destroy data.

Attack Trees

As an alternative to the above models, attack trees provide a way of organizing and representing potential threats against a system in a tree structure. Brainstorming sessions using a whiteboard are very helpful exercises and creating separate attack trees for each project is a good way to organize your thoughts. There is no one way to create an attack tree, however for a starting frame of reference, Figure 3.11 shows a generic structure of a sample attack tree.

At the top of the tree, the *root node* can be viewed from different perspectives depending on your preference such as a *verified incident* that has been researched by an analyst or the *goal* of the attacker for a specific threat. For example, say the goal of an attack is to infect a file with a virus. The next level down from the root represents the *objectives* of the attack. In our example, the goal of the attack could have two objectives: (1) execute the virus using the administrator login or (2) execute the virus as a normal user without admin privileges. The third level down in an attack tree represents the *methods* in which an attacker could carry out the attack. For example, the first method

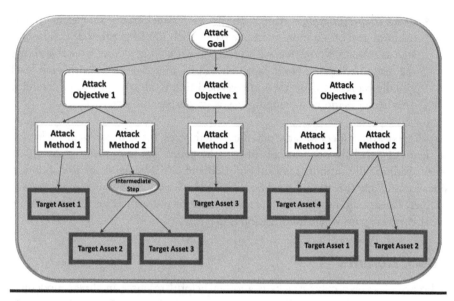

Figure 3.11   Attack tree structure.

in which the virus is run using the administrator login is that the virus exploits the root hole. Another method is that the virus is literally run by a disgruntled administrator as a malicious insider threat. The next and bottom level of an attack tree represents the *target asset* in which the attacker is ultimately working to gain access or destroy. To continue our example, one target may be that the attacker wants to infect an install package. Other targets could be other programs or files on a specific file server.

As mentioned, there is no one way to create an attack tree, however the following is a list of the basic steps involved (Shostack, 2014):

- *Decide on representation*—Business rule logic of AND/OR is used here with the idea of AND trees and OR trees. AND trees are where the state of a node (branches) depends on all the nodes below it being true. OR trees are generally used where a node is true if any of its subnodes are true. The most common types of attack trees are OR trees.
- *Create a root node*—Decide how to view the event—from an incident or problem perspective or from the perspective of an adversary's goal.
- *Create subnodes*—It is helpful to think of the first-level subnode as *objectives*, the second-level as *methods employed*, and the third-level as the *targeted assets*.
- *Consider completeness*—Determine whether your attack tree(s) are complete enough by looking for additional ways to reach the goal and asking "is there another way that could happen?"
- *Prune the tree*—In this step, review each node in the tree and consider whether the action would be prevented by a security control or countermeasure that has been implemented in the organization or if there are any redundancies.

■ *Check the presentation*—Aim to capture each attack tree or subtree in graphical form and on no more than one page. Each top-level subnode can be created as a root of a new tree, with a full context or "big picture" tree that shows the overall relations. Node labels should be in active terms with each level visually represented on the same vertical line to clearly show the hierarchy. This becomes more challenging as you go deeper into a tree.

For those of you leaders who are detail-oriented types, the addition of any effective mitigations that have been successfully implemented can be shown on the attack trees. Additionally, authors Howard and LeBlanc (2003) suggest the use of dotted lines for unlikely threats and solid lines for likely threats. The visual representation of likely threats and mitigations already in place (or not) can be very helpful in the event there are gaps where a mitigation is needed.

Attack Libraries

A library of known cyberattacks is another helpful resource in threat modeling. A *Common Attack Pattern Enumeration and Classification* (CAPEC), managed by the MITRE Corporation, was established by the United States Computer Emergency Readiness Team (USCERT) at the Department of Homeland Security (DHS). The objective of the CAPEC effort is to provide a publicly available community resource for identifying and understanding attack patterns generated from in-depth analysis of specific real-world exploit examples and solutions (CAPEC, 2018). The site includes a catalog of common attack patterns and comprehensive schema for distributing and sharing information about them. Attack patterns are a useful tool to build cyber-enabled capabilities that can be effectively used to communicate common ideas with others. They, along with security requirements, misuse/abuse cases, attack trees, and knowledge of common weaknesses and vulnerabilities are all parts of an overall formal threat model. A sample threat model is provided at the end of this chapter.

As of this writing, CAPEC has identified 566 attack patterns and can be viewed from two views: (1) by *mechanisms* of attack and (2) by *domains* of attack (CAPEC, 2018). *Mechanisms* of attack represent the different techniques used to attack a system, but do not include the consequences or the goals of the attack. There are nine categories identified to group patterns that share a common characteristic and within each category details of specific methodologies and techniques are shared (CAPEC, 2018). The following are the nine highest-level categories of the *mechanisms* attack patterns:

1. Collect and analyze information
2. Unexpected items
3. Engage in deceptive interactions
4. Manipulate timing and state

5. Abuse existing functionality
6. Employ probabilistic techniques
7. Subvert access control
8. Manipulate data structures
9. Manipulate system resources

Domains of attack organizes attack patterns hierarchically based on the attack domain. There are six domains of attack:

1. Social engineering
2. Supply chain
3. Communications
4. Software
5. Physical security
6. Hardware

For individuals new to cybersecurity enumeration, CAPEC entries include quite a bit of helpful information such as an assessment of what resources are required to carry out the attack, skills and knowledge required, and solutions and mitigations that can be implemented. For each attack pattern, a variety of detailed information is provided such as

- a summary of the attack/description
- the attack steps
- attack prerequisites
- typical severity (low-high)
- typical likelihood of exploit (low-high)
- methods of attack
- examples-instances
- attacker skills or knowledge required
- resources required
- indicators-warnings of attack
- obfuscation techniques
- solutions and mitigations
- attack motivation
- injection vector
- payload
- activation zone
- payload activation impact
- related weaknesses
- related attack patterns
- relevant security requirements
- related security principles

- related guidelines
- purposes
- CIA impact (low-high)
- technical context
- references
- content history to include submitter, date, comments, and when modified

For threat modelers, the search capability of the site can provide ideas and solutions when you're trying to identify possible threats and how to deal with them. If you've come to the end of your brainstorming sessions and want to check for completeness in your attack trees, searching the CAPEC site is a good place to conduct research.

Components of a Sample Threat Model

Ideally threat modeling should be conducted early on in the system/software development life cycle (SDLC) of any project. However, many software developers and network architects do not generally perform threat modeling, and security is often an afterthought or certainly comes too late in the life cycle. The inclusion of threat modeling in the SDLC can help ensure that systems and applications are being designed with security built-in from the very beginning (OWASP, 2018). A formal threat model can be organized into the following 13 sections:

1. Threat model project information
2. External dependencies
3. Entry points
4. Target assets
5. Trust levels/permissions
6. DFDs
7. Threat categorization
8. Security controls
9. Threat analysis (STRIDE)
10. Use/misuse case diagrams
11. Risk ranking of threats (DREAD)
12. Countermeasure currently in place
13. Documented risk mitigation strategies

The following are detailed examples of each section listed above in the context of a software development threat modeling project, and have been derived from OWASP's approach to analyzing software application security (OWASP, 2018). They can also be used as checklists during the design phase to consider potential threats and plans for mitigation.

Threat Model Project Name	
Project Version:	1.0
Project Description:	This threat model is for an online, web-based Bill Payment and Presentment System that will be accessed remotely by customers with account balances, site administrators, and customer service representatives. There will be four users of the application: 1. End users-individuals who are customers and have balances on their accounts 2. Employees of organization with elevated privileges to administer site 3. 3rd party site administrators and software developers of site 4. 3rd party Customer Service Representatives Customers (end-users) will be able to log in and search invoices, access payment history, make payments on open invoices, sign up for autopay, input payment type information. Site administrators will have additional functionality to key in business rules and other administrative functionality.
Created Date:	21 Mar 2017
Revision Date:	21 June 2018
Document Owner/Created by:	Anne K. (CISO)
Participants:	Laura B. (Project Mgr.), Dennis B. (InfoSec), Mark K. (InfoSec), Paula B. (Developer), Lisa M. (Net. Architect), Kevin C. (IT)
Reviewers:	CISO, Director of IT, InfoSec Team Lead

Figure 3.12   Section 1: Threat model example.

Section 1: Threat Model Project Information

This section documents information relating to the threat model and must include the following, also shown in Figure 3.12:

- *Threat model project name*—It is helpful to give distinct names for each project so that when discussions occur, everyone is speaking the same language. This helps cut down on any confusion, especially if there are multiple projects being worked on concurrently.
- *Threat model project version number*—From a project management perspective, each time a change is made to the threat model, a new version number should be updated.
- Description of the project
- Created date
- Revision date
- Document owner/Author/Created by:
- Participants
- *Reviewer (optional)*—Some organizations have reviewers and/or require project approval signatures.

Section 2: External Dependencies

External dependencies are items external to the code of the application that may pose a threat to the application. These items are typically still within the control of

External Dependencies	
ID Number:	Description
1.	The bill payment and presentment website will run on a Linux server running Apache. This server will be hardened as per Company ABC, who is hosting the website, server hardening standards. This includes the application of the latest operating system and application security patches. The standards are in compliance with NIST, PCI, and HIPAA regulations.
2.	The database server will be a scalable, NoSQL key-value database using Riak. This server will be hardened as per Company ABC, who is hosting the website, server hardening standards. This includes the application of the latest operating system and application security patches. The standards are in compliance with NIST, PCI, and HIPAA regulations.
3.	The connection between the Web Server and the database server will be over a private network.
4.	The Web Server is behind a firewall and the only communication available is TLS.

Figure 3.13   Section 2: External dependencies example.

the organization, but possibly not within the control of the development team. The first area to look at when investigating external dependencies is how the application will be deployed in a production environment and what the requirements are surrounding this. This involves looking at how the application is or is not intended to be run. For example, if the application is expected to be run on a server that has been hardened to the organization's hardening standard and is expected to sit behind a firewall, then this information should be documented in the external dependencies section. External dependencies should be documented as follows and shown in Figure 3.13 (OWASP, 2018):

1. *Identification number*—Assign a unique ID number to the external dependency.
2. *Description* of the external dependency

Section 3: Entry Points

Entry points define the interfaces through which potential attackers can interact with the application or supply it with data. In order for a potential attacker to attack an application, entry points must exist. Entry points in an application can be layered, for example, each web page in a web application may contain multiple entry points. Entry points, shown in Figure 3.14, should be documented as follows:

- *Identification number*—A unique ID number assigned to the entry point; this will be used to cross-reference the entry point with any threats or vulnerabilities that are identified.
- *Name*—A descriptive name identifying the entry point and its purpose
- *Description*—A textual description detailing the interaction or processing that occurs at the entry point

Entry Points			
ID Number:	**Name**	**Description**	**Trust Levels**
1	HTTPS Port	The bill payment & presentment website will only be accessible via TLS. All pages within the website are layered on this entry point.	(1) Guest Web User (2) User with Valid Login Credentials (3) User with Invalid Login Credentials (4) Administrator (5) Customer Service Reps
1.1	Bill Payment & Presentment Main Page	The splash page for the bill payment & presentment website is the entry point for all users.	(1) Guest Web User (2) User with Valid Login Credentials (3) User with Invalid Login Credentials (4) Administrator (5) Customer Service Reps
1.2	Login Page	Administrator, end-users, and customer service representatives must log in to the bill payment & presentment website before they can carry out any of the use cases.	(1) User with Valid Login Credentials (2) Customer Service Reps (3) Administrator
1.2.1	Login Function	The login function accepts user supplied credentials and compares them with those in the database.	(1) User with Valid Login Credentials (2) User with Invalid Login Credentials (3) Administrator (4) Customer Service Reps
1.3	Search Entry Page	The page used to enter a search query.	(1) User with Valid Login Credentials (2) Administrator (3) Customer Service Reps

Figure 3.14   Section 3: Entry points example.

- *Trust levels*—The level of access required at the entry point is documented here. These will be cross-referenced with the trusts levels defined later in the document.

Section 4: Target Assets

Target assets are those which the attacker is interested in and can be both physical and abstract assets. Examples of physical assets are databases with customer information, employee data, medical records, or financial information. Abstract assets are defined as the reputation of the organization, patents, trademarks, and intellectual property. Assets are documented in the threat model as follows and also shown in Figure 3.15:

- *Identification number*—A unique ID is assigned to identify each asset. This will be used to cross-reference the asset with any threats or vulnerabilities that are identified.
- *Name*—A descriptive name that clearly identifies the asset
- *Description*—A textual description of what the asset is and why it needs to be protected
- *Trust levels*—The level of access required to access the entry point is documented here. These will be cross-referenced with the trust levels defined in the next step.

ID Number:	Name	Description	Trust Levels
		Target Assets	
1	***Bill Payment & Presentment Application Users***	*Assets relating to the end users of the Bill Payment & Presentation system.*	
1.1	User Login Details	The login credentials of an end user, customer service representative, or site administrator of the bill payment & presentment site.	(1) User with Valid Login Credentials (2) Database Server Administrator (3) Site Administrator (4) Customer Service Reps
1.3	Personal Data	The bill payment & presentment website will store personal information such as invoices and payment data of the customers.	(1) User with Valid Login Credentials (2) Database Server Administrator (3) Site Administrator (4) Customer Service Reps (5) Web Server User Process
2	***System***	*Assets relating to the underlying system.*	
2.1	Availability of Bill Payment & Presentment Website	The Bill Payment & Presentation Application Webserver should be available 24/7 and accessed by all users with valid credentials.	(2) Database Server Administrator (6) Network Administrator
2.2	Ability to Execute Code as a Web Server User	This is the ability to execute source code on the web server as a web server user.	(6) Network Administrator (5) Web Server User Process
2.3	Ability to Execute SQL as a Database Read User	This is the ability to execute SQL select queries on the database, and thus retrieve any information stored within the Bill Payment & Presentment database.	(2) Database Server Administrator (7) Database Read User (8) Database Read/Write User
2.4	Ability to Execute SQL as a Database Read/Write User	This is the ability to execute SQL. Select, insert, and update queries on the database and thus have read and write access to any information stored within the Bill Payment & Presentment database.	(2) Database Server Administrator (8) Database Read/Write User
3	***Website***	*Assets relating to the Bill Payment & Presentation Website.*	
3.1	Login Session	This is the login session of a user to the Bill Payment & Presentation website. This user could be a customer, site admin, customer service rep.	(1) User with Valid Login Credentials (3) Site Administrator (4) Customer Service Reps
3.2	Access to the Database Server	Access to the database server allows the administration of the database, giving full access to the database users and all data contained within the database.	(2) Database Server Administrator
3.3	Ability to Create Users	The ability to create users would allow an individual to create new users on the system.	(2) Database Server Administrator (3) Site Administrator (4) Customer Service Reps
3.4	Ability to Audit Data	The audit data shows all auditable events and logs that occurred within the Bill Payment & Presentation application.	(2) Database Server Administrator (3) Site Administrator (4) Customer Service Reps (6) Network Administrator

Figure 3.15 Section 4: Target assets.

Trust Levels		
ID Number:	**Name**	**Description**
1	Guest Web User	A user who has connected to bill payment & presentment website but has not provided valid credentials.
2	User with Valid Login Credentials	A user who has connected to the bill payment & presentment website and has logged in using valid login credentials.
3	User with Invalid Login Credentials	A user who has connected to the bill payment & presentment website and is attempting to log in using invalid login credentials.
4	Database Server Administrator	The database server administrator has read and write access to the database that is used by the bill payment & presentment website.
5	Website Administrator	The Website administrator can configure the bill payment & presentment website.
6	Web Server User Process	This is the process application code that is executed on the web server executes and authenticates itself against the database server.
7	Database Read User	The database user account used to access the database for read access.
8	Database Read/Write User	The database user account used to access the database for read and write access.

Figure 3.16   Section 5: Trust levels.

Section 5: Trust Levels

Trust levels represent the access rights that the application will grant to external entities. The trust levels are cross-referenced with the entry points and assets and allow the definition of the access rights or privileges required at each entry point and those required to interact with each asset. Trust levels are documented in the threat model as follows and shown in Figure 3.16:

- *Identification number*—A unique number is assigned to each trust level. This is used to cross-reference the trust level with the entry points and assets.
- *Name*—A descriptive name that allows the identification of the external entities that have been granted this trust level
- *Description*—A textual description of the trust level detailing the external entity who has been granted the trust level

Section 6: Data Flow Diagrams

As mentioned earlier in the chapter, the DFDs allow for the ability to gain a better understanding of the application by providing a visual representation of how the application processes data. Shown in Figure 3.17, the focus of the DFDs is on how data moves through the application and what happens to the data as it moves. DFDs are hierarchical in structure, so they can be used to decompose the application into subsystems and lower-level subsystems. The highest-level DFD (called the context or level 0) allows for the clarification of the scope of the application being modeled. The lower-level iterations allow for the focus on the specific processes involved when processing specific data. There are a number of symbols

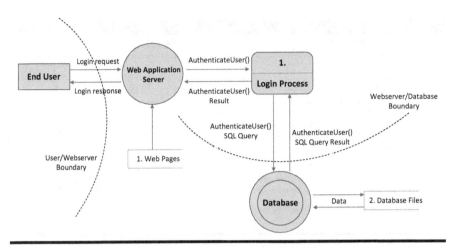

Figure 3.17   Sample DFD (repeat of Figure 3.4).

that are used in DFDs for threat modeling. Figure 3.18 shows the symbols used in a DFD.

Section 7: Threat Categorization

The first step in the determination of threats is adopting a threat categorization schema. A threat categorization schema provides a set of threat categories with corresponding examples so that threats can be systematically identified in the application in a structured and repeatable manner. A threat categorization such as STRIDE is useful in the identification of threats by classifying attacker goals such as:

- spoofing
- tampering
- repudiation
- information disclosure
- denial of service
- elevation of privilege

A threat list of generic threats organized in these categories with examples and the affected security controls is provided in the following Figure 3.19.

Section 8: Security Controls

In this section of the threat model, it is helpful to list the security controls that have already been effectively implemented and periodically reviewed, or a discussion on

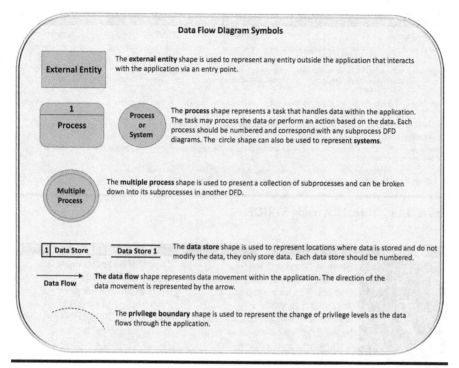

Figure 3.18   DFD symbols.

which security framework the organization has chosen to follow. There are several security control frameworks that are available for use such as the National Institute of Standards and Technology (NIST) and the Improving Critical Infrastructure Cybersecurity (CSF), which is more recent and issued by Executive Order (EO) 13636 in February 2013. NIST has one of the most rigorous and thorough frameworks and is quite comprehensive. Depending on the security requirements of the organization, security controls organized in high, medium, and low securities are available. Some security professionals who feel NIST is too much use the Payment Card Industry Data Security Standard (PCI-DSS) standards or the SAN's Top 20 Critical Security Controls. Other frameworks such as International Organization for Standardization (ISO) and Information Security Forum (ISF) are membership organizations that charge for their information. The following Figure 3.20 provides an overview of the variety of frameworks that have security control standards.

The NIST security controls are grouped into three areas: management, operational, and technical, and are widely available at no cost due to U.S. taxpayer dollars subsidizing the work. Figure 3.21 shows the security control library families for each area of the NIST security control library.

Threat List Using STRIDE Model		
Type	**Examples**	**Security Control**
Spoofing	Threat action aimed to illegally access and use another user's credentials, such as username and password.	Authentication
Tampering	Threat action aimed to maliciously change/modify persistent data such as persistent data in a database and the alteration of data in transit between two computers over an open network such as the Internet.	Integrity
Repudiation	Threat action aimed to perform illegal operations in a system that lacks the ability to trace the prohibited operations.	Non-Repudiation
Information Disclosure	Threat action to read a file that one was not granted access to or to read data in transit.	Confidentiality
Denial of Service	Threat aimed to deny access to valid users such as by making a web server temporarily unavailable or unusable.	Availability
Elevation of Privilege	Threat aimed to gain privileged access to resources for gaining unauthorized access to information or to compromise a system.	Authorization

Figure 3.19 Threat list using STRIDE.

Framework	Focus	Sponsoring Organization
COSO	Financial operations and risk management.	Committee of Sponsoring Organizations (COSO)
ITIL	Best practices for managing and delivering IT services.	Information Technology Infrastructure Library (ITIL)
ISO	International member organization focusing on IT service management, information security management, corporate governance of IT security, IT risk management, and quality management.	International Organization for Standardization (ISO)
COBIT	International governance, assessment, and management of IT security and risk management process.	Information Systems Audit and Control Association (ISACA)
NIST	IT security standards for federal agencies mandated by the Federal Information Security Management Act (FISMA).	National Institute of Standards and Technology (NIST)
CSF	Voluntary risk-based framework that focuses on IT security and risk management processes.	Presidential Executive Order 13636, Improving Critical Infrastructure Cybersecurity, dated 12 Feb 2013
ISF	International member organization focusing on IT security, governance, and managing information risk.	Information Security Forum (ISF)
PCI DSS	IT security standard for the protection of credit card account data security. Card companies include Visa, MasterCard, American Express, Discover, and Japan Credit Bureau.	Payment Card Industry (PCI) Security Standards Council
SANS Institute	Although not a framework, the widely adopted Top 20 Critical Security Controls is based on the NIST SP 800-53 control standards.	SANS Institute

Figure 3.20 IT security frameworks overview.

Within each control family, a detailed group of specific security controls are outlined and described. Figure 3.22 shows an example of the 20 individual controls that fall into the *Access Control Family* of the NIST security controls.

	Security Control CLASS	Security Control FAMILY	IDENTIFIER
1	Management	Certification, Accreditation, and Security Assessments	CA
2	Management	Planning	PL
3	Management	Risk Assessment	RA
4	Management	System and Services Acquisition	SA
5	Operational	Awareness and Training	AT
6	Operational	Configuration Management	CM
7	Operational	Contingency Planning	CP
8	Operational	Incident Response	IR
9	Operational	Maintenance	MA
10	Operational	Media Protection	MP
11	Operational	Physical and Environmental Protection	PE
12	Operational	Personnel Security	PS
13	Operational	System and Information Integrity	SI
14	Technical	Access Control	AC
15	Technical	Audit and Accountability	AU
16	Technical	Identification and Authentication	IA
17	Technical	System and Communications Protection	SC

Figure 3.21 NIST security control library families.

Security Control NUMBER		Access Control 20 Individual Controls	
1	AC-1	Access Control Policy and Procedures	The organization develops, disseminates, and periodically reviews/updates: (i) a formal, documented, control policy; and (ii) formal, documented procedures to facilitate the implementation of the access control policy and associated controls.
2	AC-2	Account Management	The organization manages information system accounts, employs automated mechanisms, terminates temporary, emergency, and inactive accounts after org-defined time; ensures actions are audited; and reviews information system accounts.
3	AC-3	Access Enforcement	The organization enforces assigned authorizations for controlling access to the system; ensures that access to security functions and information is restricted to authorized personnel in accordance with applicable policy.
4	AC-4	Information Flow Enforcement	The information system enforces assigned authorizations for controlling the flow of information within the system and between interconnected systems in accordance with applicable policy.
5	AC-5	Separation of Duties	The information system enforces separation of duties through assigned access authorizations.
6	AC-6	Least Privilege	The information system enforces the most restrictive set of rights/privileges or accesses needed by users.
7	AC-7	Unsuccessful Login Attempts	The information system enforces a limit of consecutive invalid access attempts by a user during a time period.
8	AC-8	System Use Notification	The information system displays and approved, system use notification message before granting system access.
9	AC-9	Previous Logon Notification	The information system notifies the user, the date and time of the last logon, and the number of unsuccessful logon attempts since the last successful logon.
10	AC-10	Concurrent Session Control	The information system limits the number of concurrent sessions for any user.
11	AC-11	Session Lock	The information system prevents further access to the system by initiating a session lock that remains in effect until the user reestablishes access using appropriate identification and authentication procedures.
12	AC-12	Session Termination	The information system automatically terminates a session after an org-defined time of inactivity.
13	AC-13	Supervision and Review—Access Control	The organization supervises and reviews the activities of users with respect to the enforcement and usage of information system access controls.
14	AC-14	Permitted Actions without Identification or Authentication	The organization identifies specific user actions that can be performed on the information system without identification or authentication.
15	AC-15	Automated Marking	The information system marks output using standard naming conventions to identify and special dissemination, handling, or distribution instructions.
16	AC-16	Automated Labeling	The information system appropriately labels information in storage, in process, and in transmission.
17	AC-17	Remote Access	The organization employs automated mechanisms to facilitate the monitoring and control of remote access methods.
18	AC-18	Wireless Access Restrictions	The organization establishes usage restrictions and implementation guidance for wireless technologies and documents, monitors, and controls wireless access.
19	AC-19	Access Control for Portable and Mobile Devices	The organization uses authentication and encryption to protect wireless access to the information system.
20	AC-20	Use of External Information Systems	The organization establishes usage restrictions and guidance for portable and mobile devices and employs removable hard drives or cryptography to protect information residing on portable and mobile devices.

Figure 3.22 NIST access control security control library.

Section 9: Threat Analysis

The prerequisite for the analysis of threats is the understanding of the generic definition of risk, which is the probability that a threat agent will exploit a vulnerability to cause a negative impact to the application or system. From the perspective of risk management, threat modeling is the systematic and strategic approach for identifying and enumerating threats to an application or system environment with the objective of minimizing risk and the associated impacts. Threat analysis as such is the identification of the threats to the application and/or system and involves the analysis of each aspect of the functionality and architecture and design to identify and classify potential weaknesses that could lead to an exploit.

In the earlier threat modeling steps, we have modeled the system showing data flows, trust boundaries, process components, and entry and exit points. Data flows show how data flows logically through the end-to-end of the system and allows the identification of affected components through critical points (i.e., data entering or leaving the system, storage of data) and the flow of control through these components. Trust boundaries show any location where the level of trust changes. Process components show where data is processed, such as web servers, application servers, and database servers. Entry points show where data enters the system (i.e., input fields, methods) and exit points are where it leaves the system (i.e., dynamic output, methods). Entry and exit points define a trust boundary.

Threat lists based on the STRIDE model are useful in the identification of threats in regard to the attacker goals. For example, if the threat scenario is attacking the login, would the attacker brute force the password to break the authentication? If the threat scenario is to try to elevate privileges to gain another user's privileges, would the attacker try to perform forceful browsing?

It is vital that all possible attack vectors be evaluated from the attacker's point of view. For this reason, it is also important to consider entry and exit points, since they could also allow the realization of certain kinds of threats. For example, the login page allows sending authentication credentials, and the input data accepted by an entry point has to validate for potential malicious input to exploit vulnerabilities such as structured query language (SQL) injection, cross-site scripting, and buffer overflows. Additionally, the data flow passing through that point has to be used to determine the threats to the entry points to the next components along the flow. If the following components can be regarded as critical (e.g., the hold sensitive data), that entry point can be regarded more critical as well. In an end-to-end data flow, for example, the input data (i.e., username and password) from a login page, passed on without validation, could be exploited for a SQL injection attack to manipulate a query for breaking the authentication or to modify a table in the database.

Exit points might serve as attack points to the client (e.g., XSS vulnerabilities) as well for the realization of information disclosure vulnerabilities. For example, in case of exit points from components handling confidential data (e.g., data

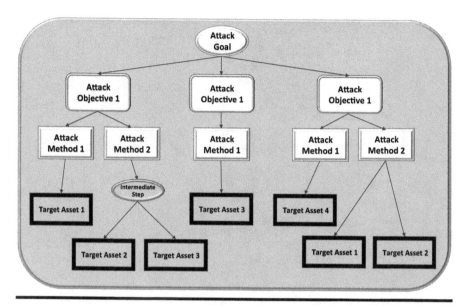

Figure 3.23   Attack tree structure (repeat of Figure 3.11).

access components), exit points lacking security controls to protect the confidentiality and integrity can lead to disclosure of such confidential information to an unauthorized user.

In many cases, threats enabled by exit points are related to the threats of the corresponding entry point. In the login example, error messages returned to the user via the exit point might allow for entry point attacks, such as account harvesting (e.g., username not found) or SQL injection (e.g., SQL exception errors).

From the defensive perspective, the identification of threats driven by security control categorization allows a threat analyst to focus on specific issues related to weaknesses (e.g., vulnerabilities) in security controls. Typically, the process of threat identification involves going through iterative cycles where initially all the possible threats in the threat list that apply to each component are evaluated.

At the next iteration, threats are further analyzed by exploring the attack paths, the root causes (e.g., vulnerabilities depicted as orange blocks) for the threat to be exploited, and the necessary mitigation controls (e.g., countermeasures depicted as green blocks). The threat attack tree structure as shown in Figure 3.23 is useful in performing such threat analysis.

Section 10: Use/Misuse Cases

Once common threats, vulnerabilities, and attacks are assessed, a more focused threat analysis should take in consideration use and *abuse cases*. By thoroughly analyzing the use cases and scenarios, weaknesses can be identified that could lead to

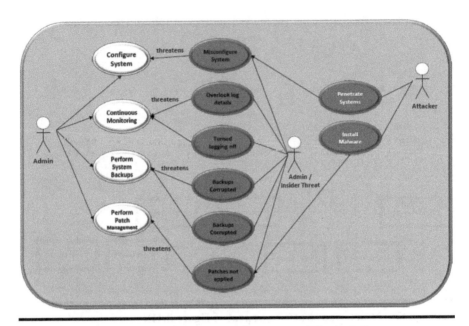

Figure 3.24   Misuse case diagram example (repeat of Figure 3.6).

the realization of a threat. Abuse cases should be identified as part of the security requirement engineering activities and illustrate how existing protective measures could be bypassed or where a lack of such protection exists. A use and misuse case diagram for authentication is shown in Figure 3.24.

Section 11: Risk Ranking of Threats Using DREAD

Threats can be ranked from the perspective of risk factors, and by determining the risk factor posed by the various identified threats, it is possible to create a prioritized list of threats to support a risk mitigation strategy to decide which threats have to be mitigated first. Different risk factors can be used to determine which threats can be ranked as *High, Medium*, or *Low* risk.

In the Microsoft DREAD threat-risk ranking model, the technical risk factors for impact are Damage and Affected Users, while the ease of exploitation factors are Reproducibility, Exploitability, and Discoverability. This risk factorization allows the assignment of values to the different influencing factors of a threat. To determine the ranking of a threat, the threat analyst has to answer basic questions for each factor of risk, for example:

- For Damage: How big would the damage be if the attack succeeded?
- For Reproducibility: How easy is it to reproduce an attack to work?
- For Exploitability: How much time, effort, and expertise are needed to exploit the threat?

■ For Affected Users: If a threat were exploited, what percentage of users would be affected?
■ For Discoverability: How easy is it for an attacker to discover this threat?

By referring to the college library website, it is possible to document sample threats related to the use cases such as:

Threat: Malicious users view confidential information of customer data in the bill presentment and payment system to include credit card and banking accounts.

1. Damage potential: Threat to reputation as well as financial and legal liability: 8
2. Reproducibility: Fully reproducible: 10
3. Exploitability: Requires to be on the same subnet or to have compromised a router: 7
4. Affected users: Affects all users: 10
5. Discoverability: Can be found out easily: 10

Overall DREAD score: $(8 + 10 + 7 + 10 + 10)/5 = 9$. In this case, having nine on a ten-point scale is certainly a high-risk threat.

Section 12: Countermeasure Currently in Place

The purpose of the countermeasure identification is to determine if there is some kind of protective measure (e.g., security control, policy measures) in place that can prevent each threat previously identified via threat analysis. Vulnerabilities are then those threats that have no countermeasures. When using STRIDE, the following threat-mitigation table, shown in Figure 3.25, can be used to identify techniques that can be employed to mitigate the threats.

Section 13: Document Risk Mitigation Strategies

It is important to document the decisions that were taken to mitigate each threat that was found. After six months or after the project team has moved on to other assignments, it is easy to forget why certain decisions were made. Some risks, once identified, can be quickly addressed with a security countermeasure or training. Other more high-impact risks will take more work and cost. The cost to mitigate a risk may not be worth the effort to upgrade the technology or systems, especially if the probability of an attack is low. Good risk management includes a full analysis of

■ the costs in monetary resources;
■ if the organization has the resources such as talent and expertise to execute the mitigation;
■ the probability of each event that may occur;
■ the timeliness in which it would take to mitigate and monitor.

Threat Mitigation List Using STRIDE Model	
Threat Type	**Mitigation Technique**
Spoofing Identity	1. Appropriate authentication 2. Protect confidential data 3. Encrypt customer data
Tampering with Data	1. Appropriate authorization 2. Use hash to encrypt 3. MAC's 4. Use digital signatures 5. Tamper resistant protocols
Repudiation	1. Use digital signatures 2. Timestamps 3. Audit trails
Information Disclosure	1. Appropriate authorization 2. Privacy-enhanced protocols 3. Encryption 4. Protect secrets 5. Don't' store secrets
Denial of Service	1. Appropriate authentication 2. Appropriate authorization 3. Filtering 4. Throttling 5. Quality of Service
Elevation of Privilege	1. Run with least privilege

Figure 3.25 Threat and mitigation security controls list using STRIDE model.

There are four strategies that can be considered as each threat is identified, they are:

1. *Risk Transfer*—This includes assigning the responsibility of risks to other parties such as contractors, suppliers, insurance firms, outsourcing, and the like. The organization and all parties are in agreement of the magnitude of the risks and who is to bear them in the event of an attack.
2. *Risk Acceptance*—This includes decisions to accept the risk of a particular threat when the cost of other risk management strategies outweigh the benefit of mitigating that threat. If the probability of an attack is low, an organization may decide to take no action and accept the possibility of an adverse event on the target asset. On the other hand, if the probability of an attack is

high and the asset is mission critical to the organization, other alternatives to mitigate the risk should be taken.

3. *Risk Avoidance*—This option is the opposite of risk acceptance and includes the elimination of risk by either terminating the service or asset to avoid any risk whatsoever. This option is usually the most expensive.

4. *Risk Limitation/Buffering*—This option is the most common, taken by organizations to limit the exposure by taking some kind of action. An organization may establish some reserve or buffer that can absorb the effects of the attack in the event of unpreparedness or not having the talent to properly monitor and secure assets. Risk limitation can also include the allocation of resources to allow for future uncertainties.

As stated, since people have a tendency to forget why decisions were made or who made them at the time, it is very beneficial to document what risk mitigation strategy was taken for each threat.

Chapter Summary

Threat modeling is something that is widely engaged in but often taken for granted. In other words, many people threat model yet they may not realize that is what they're actually doing. Threat modeling is the process of analyzing what might go wrong and to consider all the possibilities of what could happen for any given situation.

Threat modeling involves a variety of skills and knowledge about the organization such as

- an understanding of the organization's security requirements;
- knowledge of the organization's mission critical assets and the owners of the data such as department managers or application teams;
- knowledge of databases, DFDs, and where the data is stored;
- knowledge of network architecture, infrastructure components, entry points, and technologies in place to continuously monitor the network traffic;
- knowledge of what security controls have been implemented and, ideally, knowledge of security controls that have been violated or are routinely ignored;
- understanding of risk management to include risk identification, risk estimation, evaluation of risk levels and acceptance criteria, risk treatment options, and residual risk assessment.

It is critical to start at the beginning and understand the business, specifically the business requirements; essential processes; mission critical systems and data;

activities that must be performed to meet business objectives and goals; legal and governmental regulations; and customer needs/expectations. This information is then used to determine functional security requirements such as

■ access control
■ privacy
■ user data protection
■ security audits
■ trusted path/channels
■ data integrity
■ availability

There is a significant interdependency between functional security requirements, threats, and how the organization chooses to mitigate those threats. As you model threats, decisions will have to be made as to whether the cost to address the threat exceeds the benefits. Some threats may not be worth addressing and understanding the organization's business requirements and strategic plans will help guide those decisions. It is important that whatever decisions are made to address threats, they are fully documented and communicated.

There are two common approaches to structuring a threat model: (1) focusing on critical assets and (2) focusing on software applications, both in stages of development and software applications that are in production. A third, but less common approach, focuses on potential human-centered attackers and requires research and experience to develop profiles of attackers.

Once you understand the organization's security requirements and the variety of potential threat agents, select a threat modeling approach that suits your organization; armed with current system diagrams, you can then begin to find threats. There are several models or techniques that can aid in finding risks and provide structure for your threat model. They are STRIDE, DREAD, OWASP's Top Ten Project, attack trees, and attack libraries.

Ideally, threat modeling should be conducted early on in the SDLC of any project. However, many software developers and network architects do not generally perform threat modeling, and security is often an afterthought or certainly comes too late in the life cycle. The inclusion of threat modeling in the SDLC can help ensure that systems and applications are being designed with security built-in from the very beginning (OWASP, 2018). A formal threat model can be organized into the following 13 sections:

1. Threat model project information
2. External dependencies
3. Entry points
4. Target assets
5. Trust levels/permissions

6. DFDs
7. Threat categorization
8. Security controls
9. Threat analysis (STRIDE)
10. Use/misuse case diagrams
11. Risk ranking of threats (DREAD)
12. Countermeasure currently in place
13. Documented risk mitigation strategies

References

Aseef, N., Davis, P., Mittal, M., Sedky, K., and Tolba, A. (2005), Cyber-criminal activity analysis. *White Paper.*

CAPEC. (2018), *Common Attack Pattern Enumeration and Classification: A Community Resource for Identifying and Understanding Attacks,* The MITRE Corporation, McLean. Downloaded from the World Wide Web on June 21, 2018 at https://capec.mitre.org/about/index.html.

Casey, T. (2007), Threat agent library helps identify information security risks. *White Paper*: Intel Corporation.

Cooper, A., Reimann, R., and Cronin, D. (2007), *About Face 3: The Essentials of Interaction Design,* Wiley Publishing, Inc., Indianapolis, IN.

Cooper, A., Reimann, R., Cronin, D., and Noessel, C. (2014), *About Face: The Essentials of Interaction Design,* 4th Ed. Wiley Publishing, Inc., Indianapolis, IN.

Howard, M. and LeBlanc, D. (2003), *Writing Secure Code,* 1st Ed. Microsoft Press, Redmond.

Howard, M. and LeBlanc, D. (2009), *Writing Secure Code,* 2nd Ed. Microsoft Press, Redmond.

OverStack. (2018), Security/OSSA-Metrics. Downloaded from the World Wide Web on June 18, 2018 at https://wiki.openstack.org/wiki/Security/OSSA-Metrics.

OWASP. (2017), OWASP Top 10–2017: The ten most critical web application security risks. Downloaded from the World Wide Web on June 18, 2018 at https://www.owasp.org/images/7/72/OWASP_Top_10-2017_%28en%29.pdf.pdf.

OWASP. (2018), CRV2 AppThreatModeling. Downloaded from the World Wide Web on Jan 19, 2018 at https://www.owasp.org/index.php/CRV2_AppThreatModeling.

Shostack, A. (2014), *Threat Modeling: Designing for Security,* John Wiley & Sons, Inc., Indianapolis, IN.

Verizon. (2018), Data breach investigations report, Verizon Corporation, 2018.

Chapter 4

Prioritization of Assets and Establishing a Plan for Resilient Change

Following this chapter, the reader will understand

1. the processes for eliciting resilience requirements at the organizational and service levels;
2. the proper procedures for prioritizing assets for each asset type based on what is known about organizational and service level requirements, and the vulnerabilities each asset possesses;
3. a planned approach to implementing and managing resilience configuration sustainment;
4. effective approaches for monitoring change that affect the resilience strategy set forth by the organization.

Setting Prioritization into Context

Subject upon an organization having identified and baselined its assets and further characterizing the threat environment, the baselined assets should then be ranked in terms of their exposure to the known threats and vulnerabilities, cost/benefit of mitigation, interdependencies between other assets within the organization and supply chain partners, and be consistent with an understanding of the future direction of the organization, also shown in Figure 4.1.

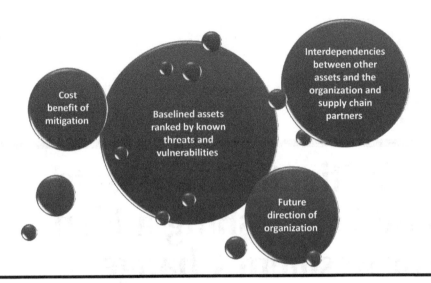

Figure 4.1 Setting prioritization into context.

As we have emphasized in each of the previous chapters, "assets" comprise all the people, information, software, technologies, and facilities required to achieve the organizational mission and objectives.

It should be made clear that through asset management, assets are matched to how they contribute to critical organizational services. From that analysis, a baseline is established through the existence of an asset inventory. That baseline then provides a basis from which vulnerabilities should be identified. With knowledge of threats and vulnerabilities, the organization can move forward in developing a clear set of resilience requirements. Further, each of the assets within the asset inventory can be ranked and documented based on the criticality of each asset in satisfying the established resilience requirements and assuring the accomplishment of the organization's mission, vision, values, and purposes.

Unfortunately, requirements elicitation and ranking can often turn into a political free-for-all, where each stakeholder attempts to enforce its own agenda. Obviously, this cannot be allowed to happen if the eventual architectural solution is going to be truly resilient. Therefore, criticality must be understood based on a clear map of functions and dependencies, which are referenced in an objective and rational way to the mission and goals of the organization. This chapter will discuss how an asset can only be labeled "critical" if it provably underwrites some aspect of the organization's core functionality.

This chapter also introduces the process for creating a rigorous set of protection requirements for just those assets that directly enable the organizational mission. Rigor, in this case, is defined as the ability to resist any known or conceivable method of attack. Through prioritized resilience requirements, relevant stakeholders are assigned to supervise and maintain each asset, and effective communication linkages are established between those stakeholders and documented.

Resilience Requirements Elicitation and Definition

Guided by the frameworks that prescribe the processes of the standard system development life cycle, it is generally understood that decisive approaches to identification and documentation of functional and nonfunctional requirements is a requisite for the overall quality of the system under development. System analysts spend a considerable amount of time identifying stakeholders, eliciting requirements, developing models, and documenting analysis results in a requirements specification that is then validated and used as input for the succeeding phase of the life cycle. What is significant here is that the requirements being sought are generally appropriate to the level of abstraction from which the system or subsystem exists within the organization. While the argument can be made that the process of eliciting and defining requirements for cyber resilience does still consider functional and nonfunctional characteristics, it does so at the higher-level abstraction of the organization as compared to the more granular system level. Before going further, however, it might be beneficial to define "requirement" from a cyber resilience perspective.

A cyber resilience requirement consists of some characteristic, condition, or capability that an organizational asset must have in order for it to remain justifiable in terms of the support it provides to a critical service of the organization. Generally, cyber resilience requirements have a relationship to the cybersecurity controls implemented within the organization through their underlying focus on established measures of confidentiality, integrity, and availability (CIA). While we have emphasized the departure of cyber resilience to the principles of cybersecurity, these CIA properties normally associated with information assets are also applicable to the other asset types of people, software, technology, and facilities, which serve as a basis from which cyber resilience management is performed.

As we emphasized in Chapter 1, cyber resilience is based on a well-defined set of resilience requirements that provide the foundation for protecting assets from threats and provide capability for maintaining their effectiveness to the extent that they can perform as intended in support of critical services. Put differently, resilience requirements become a part of an asset's identity in much the same way as the asset's definition, owner, custodian, and value contribute to the organization's ability to achieve established objectives (sometimes reaching beyond organizational boundaries into the processes of supply chain partners) and are characterized through asset management.

Further, in the same way as system requirements are a precursor to successful design and implementation, cyber resilience requirements drive many of the organizational level processes, such as cyber resilience management. Cyber resilience management, as it relates to requirements elicitation and definition, for example, provides the capability for the organization to perform the prescribed activities and tasks that establish resilience requirements at the organizational, service, and asset levels. Likewise, resilience requirements provide the basis for developing protection strategies and security controls in addition to providing the capacity for developing business continuity plans for critical services and their supporting assets.

The importance of requirements for the resilience management system cannot be overstated. Resilience requirements embody the strategic objectives, risk appetite, critical success factors (CSFs), and operational constraints of the organization. They represent the alignment factor that ties practice-level activities performed in security and business continuity to what must be accomplished at the service and asset levels in order to move the organization toward fulfilling its mission.

Of course, the focus of this book is on cyber resilience from an asset perspective. It is important to note, however, that depending on the organization's priorities, three types of cyber resilience requirements may be sought: organization, service, and asset, as shown in Figure 4.2.

- **Organization**: Organizational cyber-resilient requirements are largely based on organizational level needs, expectations, and constraints. What is significant about these requirements is that they have an impact on most, if not all, operations. Many organizations will take a rules-based approach to define requirements at this level. Although the rules that must be applied are largely out of the organization's control. For example, most industries are held accountable to federal, state, and local laws in addition to industry-specific regulations. Examples include Health Insurance Portability and Accountability Act (HIPAA) for health care institutions, Sarbanes–Oxley

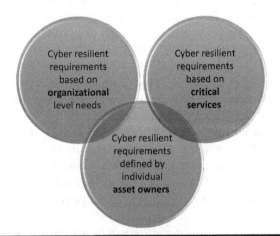

Figure 4.2 Three types of cyber resilience requirements.

for investment companies, and Gramm–Leach–Bliley for financial institutions. Each act has its own prescribed set of rules that in turn have an effect on enterprise cyber resilience requirements.

▪ **Service**: Service cyber-resilient requirements ensure that the resilience needs of each critical service within the organization is met in order to achieve the organization's mission and objectives. However, it is important to remember the direct correlation between a service's ability to satisfy the established mission and objectives and the asset resilience requirements that support that service. Therefore, service requirements must reflect and be consistent with the cyber resilience requirements of assets that support them. It is not uncommon to think of service requirements from the perspective of its availability and recoverability. Although, availability and recoverability are mainly driven by that service's ability to meet the CIA requirements of people, software, information, technology, and facilities.

▪ **Asset**: Asset cyber-resilient requirements are defined by each individual asset owner. The requirement's purpose is to establish the mechanisms necessary to protect and provide the support that asset needs in its ability to provide the capabilities each associated critical service requires to satisfy organizational objectives. Organizations must also consider that assets often have conflicting requirements across multiple services (e.g., a software application may, and often does, support more than one service). In such circumstances, proper measures must be taken to ensure that the asset is capable of providing adequate resilience to satisfy the organization's resilience priorities.

Depending upon the size of the organization, there is a plethora of approaches to resilience requirements elicitation. The approach taken is indicative of which activities of requirement elicitation are performed at the appropriate organizational level. For example, strategic planning and board policy may dictate organization-level requirements, as would interviewing of senior managers. Middle management and other critical service stakeholders will likely elicit service-level requirements. And, asset-level requirements are more likely to be determined through the organization's risk assessment process and business process reengineering efforts with influence from the assets identified by the owner and custodians. The key is to have conversations with those individuals that are familiar with how each asset contributes to or affects the services in some way in order to have a clear perception of the resilience requirements for that asset.

What should be evident from the discussions presented in this and past chapters is that a bottom/up approach to cyber resilience of organizational assets is unproductive. Rather, a decisive understanding of requirements at the organizational and service levels provides drivers the ability to consider the output of vulnerability assessment together with the categorized assets in the asset inventory in order to develop a clear set of asset resilience requirements. With that said, all resilience requirements (at all three levels) must be analyzed for conflicts and dependencies, and must provide adequate support for the organization's goals, objectives, and CSFs. In absence of that consideration, controls implemented to protect and sustain assets and services will not align with what the organization needs to accomplish in order to remain a contributor to its industry.

Therefore, working from the top—down, elicitation and definition of resilience requirements include the following generic outcomes and are shown in Figure 4.3.

- A reevaluation of the organization's mission and objectives while making them available to be used as the foundation for setting resilience requirements
- A clear and decisive set of organizational-level resilience requirements that have been communicated throughout the organization
- A clear and decisive set of service-level resilience requirements that directly support the defined organizational-level resilience requirements, and have been communicated throughout the organization
- A clear and decisive set of asset resilience requirements that directly support the defined service-level resilience requirements, and have been communicated throughout the organization

In order to accomplish those established outcomes, the resilience requirements elicitation and definition process should include the activities provided in Figure 4.4.

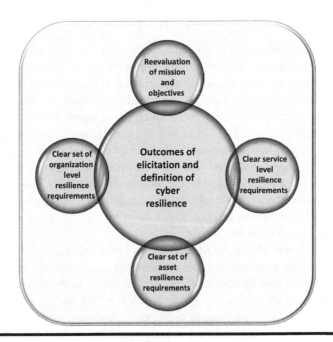

Figure 4.3 Four outcomes of elicitation and definition.

Resilience Requirements Elicitation and Definition Process		
Step 1	Identify Organizational Requirements	Resilience needs are identified at the organizational-level. These are the requirements imposed on all functions and activities within the organization.
Step 2	Identify Service Requirements	Resilience requirements to support critical services are established, in addition to the resilience requirements of the assets that support those services.
Step 3	Establish Asset Resilience Requirement Priorities	Based on the asset resilience requirements produced by vulnerability and risk assessment, assets are prioritized for control implementation.
Step 4	Analyze and Validate Requirements	The requirements are analyzed and validated to ensure appropriate asset support necessary to sustain critical services.

Figure 4.4   Resilience requirements elicitation and definition process.

Identify Organizational Resilience Requirements

At first glance, there may be a tendency to recall the activities performed within asset management at the organizational level and conclude redundancy of what is accomplished through the identification of organizational resilience requirements. To the contrary, goals of asset management and resilience requirements are very different. Through asset management the organization is focused on understanding the established mission and objectives in order to articulate the critical functions that satisfy that mission and objectives. Organizational resilience is "the ability of an organization to anticipate, prepare for, respond and adapt to incremental change and sudden disruptions in order to survive and prosper" (BSI, 2014). Put differently, at this highest level of resilience requirement abstraction, the organization is tasked with identifying the requirements that achieve "mission assurance."

Often, mission assurance is mistaken for quality assurance and used in the wrong context. Quality assurance is comprised of those activities and tasks that provide a qualifiable and quantifiable approach to ensuring a predefined level of product or service quality and can be achieved through the combination of business processes.

However, mission assurance includes the combined application of system engineering, risk management, quality, and management principles to achieve success of a design, development, testing, deployment, and operation process. Mission assurance is the measure by which an organization, supply chain affiliate, and other business partners are able to meet the criteria set forth by the mission, with an ultimate goal to create a state of resilience that supports the continuation of an organization's critical business processes, protects its employees, assets, services, and functions, and appropriately satisfies the mission and established organizational objectives.

Therefore, organizational resilience requirements meet the criteria set forth by mission assurance by determining what is needed to satisfy that protocol by articulating organizational-level needs, expectations, and constraints into a set of requirements that affect nearly all aspects of an organization's operations. It requires the organization to think broadly about the necessities (both tangible and intangible) that must remain functional and available in order to sustain business continuity. For example, a specific example of an organizational requirement for a health care institution is that all health-related information that is covered by HIPAA regulations must be kept confidential to health workers and patients. As requirements are identified at the organizational level, they are articulated to the services that have an impact on the organization's ability to meet those requirements. Later, identified services are further articulated to the assets that support them, providing a conclusive approach to ensuring cyber resilience.

Because of the external implications associated with building a coherent set of organizational resilience requirements, it is best to take an outside-in approach to proceeding through requirements elicitation that begins with understanding the applicable legal, statutory, and regulatory obligations facing the organization. In other words, organizations must begin by looking for requirements outside of their control.

Depending upon the industry, country, or the state of operation for US entities, an organization is obligated to a plethora of restrictions. Each of these restrictions must be understood and associated organizational resilience requirements developed. Figure 4.5 provides a sample of restrictions and appropriate requirements.

The preceding list of regulations and associated resilience requirements is not exhaustive by any means. The operations performed within each business unit of the organization will drive the necessity for comprehensive approaches to begin the establishment of a resilience requirement's baseline by the awareness of the standard protocols set forth by federal, state, and local governments in addition to industry accrediting bodies.

While it may be a time consuming but fairly straightforward task of identifying directly the statutory, legal, and regulatory drivers affecting an organization's operation, consideration must also be made to those drivers that are applied to the organization's supply chain partners and contractors. As much as common cybersecurity principles include the necessity for organizations to implement security controls that satisfy inclusiveness of supply chain other third-party information systems, there is a need for awareness of the legal and regulatory obligations those partners are bound to. This awareness may (and in many cases does) lead to additional resilience requirements from which the organization realistically has no control.

The implications of supply chain relationships on the ability of establishing a concrete set of organizational resilience requirements go beyond just the need to understand their industry standards and legal obligations. Given the dependency

Sample of Legal and Regulatory Protocols with Associated Requirements	
Sarbanes–Oxley Act (SOX) **Who is affected:** U.S. public company boards, management and public accounting firms	
Availability of Studies and Reports	All financial related studies and reports will be kept and be made available in accordance with industry regulation
Corporate Tax Returns	All corporate tax returns will be kept and made available for the period defined industry regulation
Payment Card Industry Data Security Standard (PCI DSS) **Who is affected:** Retailers, credit card companies, anyone handling credit card data.	
Protect Cardholder Data	All credit customer data will be protected in accordance with PCI DSS
Maintain an Information Security Policy	Security Policy will be implemented and maintained through configuration management change control.
Federal Information Security Management Act (FISMA) **Who is affected:** Federal agencies	
Security awareness training for personnel	All personnel will be properly trained in the security policy implications of their work area.
Periodic risk assessments	All organizational assets will be assessed for risk based on a predefined schedule set forth in the security policy
Massachusetts 201 CMR 17 **Who is affected:** Businesses that collect and retain personal information of Massachusetts residents in connection with the provision of goods and services or for the purpose of employment	
Selection of third-party service providers that can properly safeguard personal information	Prior to the selection of third-party service providers, all organizational units will assess the electronic and non-electronic capabilities for protecting personal information of state residents.
Limits on the collection of data to the minimum required for the intended purpose	All data collection of state residents will be rigorously evaluated for appropriateness of use and storage

Figure 4.5   Sample of legal and regulatory protocols with associated requirements.

that an organization has on its supply chain partners to utilize services and share assets for the purpose of meeting business objectives, those services and assets that cross organizational boundaries must be evaluated for the possibility that they will warrant resilience requirement consideration. The problem faced by organizations, however, is the limited information supply chain partners and third-party providers are willing to share about their services and assets. Often, an organization must rely on the content available within contracts (Service Level Agreements, Blanket Purchase Agreements, Memorandums of Understanding, and Interconnection Security Agreements), information technology (IT) system documentation, and other business relationship documentation in order to assess criteria that will assist in identifying constraints that will impact the establishment of resilience requirements at the organizational level.

Internal elicitation of organizational resilience requirements is much easier than reaching beyond the organizational boundaries. This is due to the amount of control an organization has over its own resources and related documentation. The concern for management is generally the unavailability of necessary documentation or lack of knowledge about its location. If resilience requirements are done as a follow-on to the processes of asset management, much of the information needed to elicit internal organizational resilience requirements will be readily available. At a minimum, the following items should be gathered:

- strategic objectives that must be supported by all business functions (generally provided within a strategic plan)
- the organization's mission/vision statement and documentation that provides insight into its interpretation
- any organizational policies that provide definition of acceptable behaviors across the organization. For an educational institution, this may entail the policies associated with keeping student data confidential

Too much documentation will serve the elicitation task more effectively than not enough. Successful elicitation at this level will produce and communicate a set of organizational requirements that affect all organizational functions and all business models. This is accomplished through the identification of resilience requirements that satisfy

- organizational strategy, CSFs, and adopted policies;
- identified business constraints;
- and, the identified business requirements for processing, transmitting, and storing information that directly supports and is deemed critical to business operations.

Identify Service Resilience Requirements

Common knowledge of business principles suggests that an organization's ability to achieve an adopted mission and defined strategic objectives is through the capacity through which the services of that organization are provided. The same is true of achieving organizational resilience requirements. Those requirements are achieved through the consistent and efficient performance of services and the ability to meet the resilience requirements of those services. Further, there is a dependency of the appropriate functionality and availability of assets to meet the resilience needs of each service in order to ultimately achieve organizational mission assurance objectives. As we discussed in a previous chapter and shown in Figure 4.6, those assets consist of:

- People: Manage and perform the human contribution for each service
- Information: Provides viable input to and output from each service
- Software: Provides the means by which information support for each service is adequately processed and stored; it also contributes to the effective functionality of the technology required by the service.

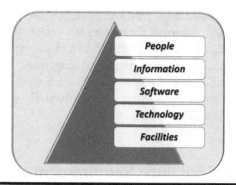

Figure 4.6   Organizational assets.

- Technology: Provides the industrial, manufacturing, telecommunications, and business system support necessary for each service to achieve its intended objectives
- Facilities: Extend the capability of the organization to place assets throughout the organization in locations providing effective performance of service objectives.

It likely comes as no surprise that a statement be made that management of each individual service is most likely to provide the information needed to determine service-level resilience requirements. To that extent, however, two important points need to be made.

One, in much the same way that organizations must consider the relationships they have with supply chain partners in order to thoroughly understand the dependencies that drive certain resilience requirements at the organizational level, careful analysis must be performed to determine dependencies among functions performed within and between organizational services. In that regard, the value-chain principle of service interoperability must be considered when identifying service-level resilience requirements. As a starting point for identifying the dependent nature of services, organizations may choose to use an elicitation technique taken from the processes set forth by recent IT methodologies. Collaboratively, management from interrelated services conduct a series of joint meetings in which the agendas are based on communication resulting in the definition of service requirements.

Two, identification of service requirements is driven by the requirements that have been identified of associated assets. Therefore, the process for elicitation of service-level resilience requirement should follow the steps as follows:

1. Identify the assets associated with each service.
2. Determine how each asset contributes to the service's ability to achieve established objectives.
3. Define the asset resilience requirements.

Thus, through the identification and definition of asset requirements, service requirements are implied. Likewise, the process of associating asset requirements to service requirements is iterative (much like modern IT methodologies) and reevaluated based on the organization's configuration management (CM) policy. Through the combined efforts of asset and service owners, each service must have an identified set of resilience requirements that can be effectively matched to dependent (internal and external) services.

Establish Asset Resilience Requirement Priorities

It should be made clear and reiterated that by the time an organization has reached the prioritization of asset resilience requirements, complete understanding of the mission and objectives has been achieved and services have been identified that are deemed critical to the accomplishment of that mission and objectives. Next, through asset management, all assets required of those services have been identified and logged in an asset inventory. Finally, the process of establishing asset resilience requirements was the conclusive output to vulnerability and risk assessment that provided a clear perspective of the risk imposed on each asset and the necessary resilience requirements to protect that asset to the extent that it maintains its baseline functionality.

Recall from the last chapter, that one core output to the assessment process is the documented requirements that provide CIA of each critical service asset. Maintaining a level of practicality to the extent at which realistic implementation of controls that meet the requirements, each asset must be ranked based on the combination of identified risk and the dependency of the asset to critical services. Each asset type is unique in its resilience requirements, and therefore requires individualized approaches to prioritization. We explore each of those approaches in the following sections.

Prioritizing People Assets

It should be noted up-front that in determining the priorities of people assets (human resources) the motivation is not finding the number of staff to have available, rather organizations must focus on making sure that the appropriate skill sets are available to maintain resilience goals and objectives. This point is particularly significant for those individuals that are directly involved in work that affects the overall resilience of the organization, such as performing security duties, supporting business continuity activities, and managing IT operations. Based on the results of vulnerability and risk assessment, effective approaches should be applied to determining what skills the organization must demonstrate in order to meet its resilience needs, and baseline competencies must be established to ensure that the entire range of necessary skills is identified to meet those needs.

Baseline competencies may be as detailed as necessary in order to describe its required set of skills and are shown in Figure 4.7. This generally includes several layers of information, including

- role or position (security administrator, network administrator, chief information officer (CIO), analyst, programmer, etc.);
- skills (C# programming, Python, structured query language (SQL) Server database administrator (DBA), security risk assessment and mitigation, etc.);
- certifications (CISSP, C|EH, CCNA, MCSE, CAP. A+, etc.);
- aptitudes and job requirements (able to work long hours, travel, or be on call).

Based on the established baseline, the organization must determine what skills currently exist and identify skill gaps that prevent efficiency and effectiveness in managing operational resilience and add additional risk to the organization in meeting its strategic objectives.

In order to develop a concrete set of baseline competencies, the organization must perform a detailed examination of the resilience program and plan. That exam

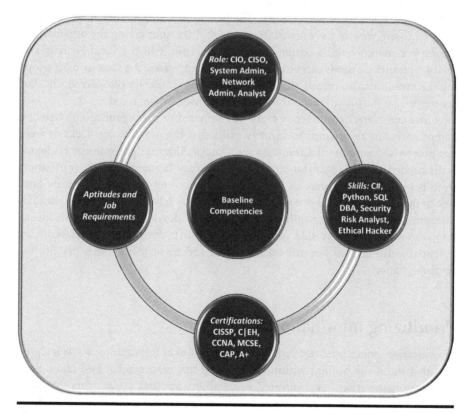

Figure 4.7 Baseline competencies.

must also extend to review job descriptions that must be developed for positions that are directly related to the organization's overall resilience program. Failure of organization to not develop job descriptions, gathering baseline competencies may be more difficult and may require an inventory of resilience positions from which a foundation for developing more extensive baselines can be created. At a minimum, the baseline competencies should be based on what the organization needs and not on what it currently has in terms of staff and skills. By determining what the organization needs, the appropriate priorities for achieving a target of sufficient staffing and skills can be established.

With a concrete set of resilience appropriate baseline of competencies in place, the next step is for the organization to develop a skills inventory, which provides a means for identifying and documenting the current skill set of the organization's human resources. This inventory is used to crossmatch the baseline to existing skills in order to diagnose resource shortages and gaps based on the organization's needs.

A skills inventory is not a job inventory or organizational chart; that information was gathered through asset management. Rather, its main purpose is to capture the skills and aptitudes of the organization's current human resources regardless of positions, roles, or responsibilities. The benefit of the inventory is to provide a comprehensive view of the organization's capabilities, thus giving the organization a complete snapshot of its current competencies from which critical analysis and review of resilience needs can be performed. In many cases, the exercise of developing the inventory also reveals the requirement for staff members to have skills that are needed by the organization that were not previously identified.

Once completed, the inventory can be compared to the organization's baseline competencies in an attempt to identify skills that the organization lacks or were not otherwise aware of. The resulting skill gap provides insight into the resilience skills needed by the organization. These skills may be keeping the organization from performing adequately in managing resilience and may result in additional risk to the organization (above and beyond what may have been identified through assessment). In identifying skill gaps, through the development of a skills inventory, the organization is able to effectively identify areas of risk that result in potentially diminished operational resilience, and in turn become a priority for implementation.

Prioritizing Information Assets

Prioritization (sometimes referred to as classification) of information assets is a process that rightfully belongs within the scope of risk management and takes place after the organization has completed vulnerability and risk assessments that have identified specific information assets (electronic or otherwise) as likely candidates for loss. This prioritization process establishes the information assets that are of most value to the organization and for which measures to protect and sustain are required. Organizations that fail to prioritize information assets risk the possibility of inadequate resilience of vital information and excessive levels of resilience for information that the organization is not as dependent on.

Given the unique nature of how each organization perceives the value (and in turn the requirement for resilience) of each piece of information moving into, within, and out of the scope of their control, different techniques are likely to be used to measure the requirements for resilience. For example, agencies of the federal government are required (through the enforcement of The Federal Information Security Management Act—FISMA) to utilize the combination of two standards: Federal Information Processing Standards (FIPS) publication 199, *Standards for Security Categorization of Federal Information and Information Systems* and the national institute for standards and technology (NIST) SP 800-60, *Guide for Mapping Types of Information and Information Systems to Security Categories.*

The FIPS 199 structure for documenting the prioritization/categorization of information is based on the potential impact baselines driven by the objectives of providing appropriate levels of resilience according to a range of risk levels. The standard defines three levels of potential impact (low, moderate, and high that can be effectively matched to the extent of vulnerability and risk of an information item. Further, FIPS 199 provides criteria for the categorization of information *types* that can be applied to all forms of information existing within the organization. Slightly modifying the FIPS 199 general syntax of categorizing information to be applicable to cyber resilience, it follows that the following can be used:

$$RC_{\text{information type}} = \left\{ (\text{confidentiality}, \text{impact}), (\text{integrity}, \text{impact}), (\text{availability}, \text{impact}) \right\}$$

where the acceptable values for potential impact are LOW, MODERATE, HIGH, or NOT APPLICABLE.

For example, suppose an educational institution manages both sensitive student information and administrative information in a registration system that has been determined must remain operationally resilient. Further, through vulnerability analysis it was determined that the potential impact (for student information) from the loss of confidentiality is *high* based on the criteria it must follow, which has been established by the Family Educational Rights and Privacy Act (FERPA); the potential impact from the loss of integrity is *high*; and the potential impact from the loss of availability is *moderate*. For administration information, the loss of confidentiality is *moderate*, the loss of integrity is *moderate*, and the loss of availability is *low*. The resulting RC for these information *types* are expressed as:

$$RC_{\text{Student Information}} = \left\{ (\text{confidentiality}, \text{HIGH}), (\text{integrity}, \text{HIGH}), \\ (\text{availability}, \text{MODERATE}) \right\}$$

and

$$RC_{\text{Administrative Information}} = \big\{ (\text{confidentiality}, \text{MODERATE}), (\text{integrity}, \text{MODERATE}),$$
$$(\text{availability}, \text{LOW}) \big\}.$$

It is important to remember that the information being categorized through this approach must have been deemed required in order to maintain the resilience of the organization. Also noteworthy by looking at the example provided above, is the close relationship that information resilience requirements have to enterprise or service requirements that have likely already been identified. In case of an educational institution, it was likely established that one of the organizational-level resilience requirements is compliancy with FERPA regulations. To that extent, it becomes necessary for the resiliency of the information pertinent to FERPA to be recognized.

The purpose of FIPS 199 is to provide the structure necessary in information prioritization that enhances clarity, through syntax, of each information type and its associated impact levels. The process for identifying information types and classifying each based on vulnerability and risk exposure is provided through NIST SP 800-60. That guideline provides a four-step process, shown in Figure 4.8, for formally prioritizing/categorizing information.

1. *Identify information types*—Develop policies regarding information identification for prioritization/categorization purposes. Following the framework presented in this book, this step would have been completed through activities associated with asset management. The deliverable for this step is a document that clearly states the organization's business and mission areas and the identification of the information types that are inputted, stored, processed, and/or outputted from each process within the organization or distributed to parties outside the organization. Additionally, this document should include the basis for the information type selection.
2. *Select the security impact* levels for the identified information types—The security impact levels can be selected either from the recommended provisional impact levels for each identified information type using the guide, NIST

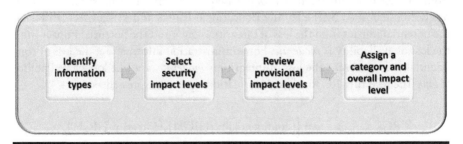

Figure 4.8 Four-step process for prioritizing information.

SP 800-60, Volume 2, Appendix C and D or from the FIPS 199 criteria for specifying the potential impact level based on resilience objectives. The deliverable for this step is a document that states the provisional impact level of CIA associated with each of the information types that must maintain resilience.

3. *Review provisional impact levels* of the information impact levels for the information types—This step also includes the *adjustment of the impact levels* as necessary based on the following considerations:
   a. CIA factors
   b. situational and operational drivers (timing, life cycle, etc.)
   c. legal or statutory reasons

   The deliverable for this step includes a document of all adjustments as well as the final impact level assigned to each information type and the rationale or justification for the adjustments.

4. *Assign a category and overall impact level*—Review the identified categorizations for the aggregate of information types and determine the categorization by identifying the highest impact level for each of the security objectives (CIA). The deliverable for this step is a document of final decisions made for the assignment of the overall information impact level based on the highest impact level for the system security objectives (CIA).

An information sensitivity categorization scheme and the corresponding information handling processes and procedures provide a way for the organization to put its mark on information assets relative to their risk tolerances and to allow for an appropriate level of corresponding handling, protection, and resilience. Failure to provide an information sensitivity categorization scheme allows for organizational staff to determine sensitivity using their own guidelines and judgment, which may vary considerably. A consistently applied sensitivity categorization scheme also ensures consistent handling of information assets across the organization and with external entities.

Prioritize Technology and Software Assets

Until now, our discussions have focused on considering technology and software assets as two separate entities. Both asset types were catalogued separately in an asset inventory through the activities of asset management, configuration identification and baselining were done separately, and both have been considered separately in order to accurately assess the risks and vulnerabilities to the organization. However, the tasks performed for the purpose of technology and software prioritization are so similar that they can be considered in combination. To be clear, when we speak of prioritizing technology assets, that can be anything from desktop computers to telecommunication systems that support supply chain relationships. Software can range from business process specific applications to operating systems or cybersecurity controls protection and detection utilities. Likewise, the focus should not simply center on the technology and software needed to keep vital services operational but also on what is needed on a broader scale to provide the underlying resilience of the entire organization.

The practice of prioritizing technology and software is not new. It's called Project Portfolio Management (PPM). What is new, however, is that organizations must integrate and prioritize cyber resilience considerations within the portfolio planning process. Traditionally, PPM has not been well understood or embraced in large organizations and sometimes is managed haphazardly. That has been rapidly changing with the outgrowth of cybersecurity threats and the newer objective of achieving cyber resilience.

Many definitions of PPM have emerged over the years. Sometimes, it is easier to describe something by explaining what it is not. PPM is not just enterprise-wide project management nor is it simply the management of projects and metrics generation across various programs and projects. PPM is the construction and management of a portfolio of projects that make a maximum contribution to an organization's overall goals and objectives (which is why it becomes so important in the discussion of prioritizing for technology and software resilience).

Organizations need to integrate cyber resilience into PPM for the following key reasons, as shown in Figure 4.9.

- PPM enables organizations to choose projects that are aligned with their overall strategy and goals.
- PPM effectively balances resource capability and project resource requirements, which eliminates inefficiencies that result from poorly staffed or overburdened teams. It also assures that resources are not being wasted while not being used on projects.
- PPM brings realism and objectivity into project planning and funding. Projects are selected because they bring value to the organization and not because of individual political agendas.
- PPM provides visibility into projects, how they are funded, and the human and financial capabilities of the organization.
- PPM follows the same principles as financial portfolio management and allows a company to maximize its return on project investment by selecting the right mix of projects.

In its simplest form, PPM can be broken down into three main components. The first component deals with building the pipeline, the second assures that the right projects are selected, and the third component deals with prioritizing the selected projects correctly. In short, PPM focuses on decision making about an organization's existing information and communication technology (ICT) products and services and those in development. It is a strategic function that aims to establish

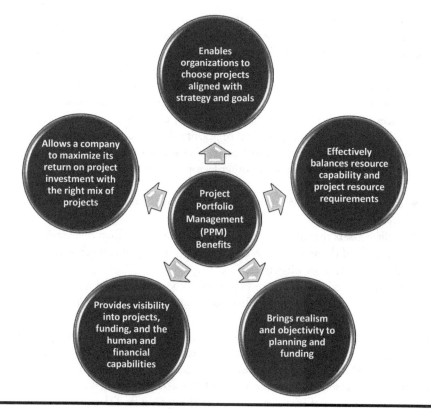

Figure 4.9 Benefits of PPM.

and maintain a balanced product portfolio that maximizes value, supports the business strategy, and makes the best use of organizational resources.

Prioritize Facility Assets

Depending upon the manager you ask, it's likely that you will hear that all facility assets for the service area they manage may be considered of high value to the organization. Realistically though, from a risk and resilience perspective, facility assets present resilience requirements that are equal in importance to the other asset types and must be prioritized. As is characteristic of the other types, prioritization of facilities is intended to identify the facilities that are of most value to the organization (based on their support for critical services), and for which protective controls and sustainability measures are required. An organization's neglect to prioritize facility assets or haphazard prioritization decision is likely to lead to inadequate resilience of assets within those facilities.

In order to appropriately assess the value of a facility and in turn provide the appropriate level of priority, it is necessary to understand the criticality of the services within each facility and the prioritization of the assets that are housed within it. Clearly, a facility that contains fewer critical services and assets, or assets that have otherwise been graded at lower levels of resilience prioritization, would in turn be graded lower. Alternatively, a facility such as a data center, which houses many critical services and high priority assets (information, technology, and software) would need to be considered at a higher level of priority than its counterparts. While management must consider the combination of service and asset priority criteria in making facility prioritization decisions, other factors should also be considered such as

■ the use of the facility asset in the general management and control of the organization;
■ facility assets that are important to supporting more than one service;
■ the value of the asset in directly supporting the organization's achievement of CSFs and strategic objectives;
■ the organization's tolerance for "pain"—the degree to which it can suffer a loss or destruction of the facility asset and continue to meet its mission.

As the prioritization process proceeds, it is not uncommon for the organization to select a subset of facility assets from its asset inventory that require the highest level of resilience. However, we cannot underestimate the possibility that a list of high priority facility assets could also result from the assessment of risk or other factors. Using a combination of resources to identify the critical facilities eliminates the risk imposed by focusing attention on just facilities contained within the asset inventory and missing facilities that may have never been inventoried.

Regardless of the approach, the goal is to have identified a list of facilities ranked based on the criticality of services and priority of assets contained within. Thus, the organization can ensure that it focuses protection and sustainability activities on facilities that have the most potential for impacting the organization if they are disrupted or destroyed.

Analyze and Validate Resilience Requirements

Once all organization and service level resilience requirements have been identified, and asset level requirements prioritized, they all must be analyzed and validated to ensure adequacy in providing the level of resilience that assets need in order to fulfill their roles in support of the organization's mission and objectives and the services that achieve them. This process is accomplished in three phases. First, a baseline understanding of how each asset meets the organizational and service level requirements is achieved. Next, each requirement (at all three levels) is analyzed to determine whether there are additional constraints that must be resolved. Lastly, asset requirements must be examined to ensure that they provide the level of resilience needed to protect and sustain that asset, along with the degree to which that accomplishes a service mission.

To establish the baseline understanding, the organization must perform an analysis and validation of the expected behaviors of each asset as those behaviors relate to the services from which they operate. By doing so, a foundation is being built that verifies that the requirements are properly aligned with organizational-level strategic objectives, in addition to the confirmation that the assets will provide the appropriate level of resilience when translated into protective controls and business continuity plans.

There are two primary reasons why the analysis of asset resilience requirements is so important and must be performed on a regular basis. One, the organization must have in place a process that provides the capability of identifying conflicts between the required functionality of the asset based on its support for business services. Put differently, for redundant assets providing the same service, an identical same level of functionality will end up being costly and it will be difficult to maintain a balance of control for resilience to be achieved for that service. Second, the organization must identify conflicts that arise when the asset is vital to more than one service requiring differing levels of resilience. This often occurs with all assets types that are shared by more than one service.

By performing a formal analysis process, the organization is also able to effectively identify requirements that cannot be met or that are inconsistent with the predetermined baseline functionality of the asset.

Each asset requirement must also be objectively validated to ensure that it supports the required functionality of the asset and its associated services. This task is performed through the creation of metrics that generate appropriate data that will assist in the identification and control of risks to protect and support assets that are introduced by requirements. The data collected will also assist in reviewing the alignment between requirements and the organization's strategic objectives capacity to be achieved through the satisfaction of resilience requirements. Additionally, it is useful in identifying any missing or inadequate asset requirements.

Establish Resiliency through Change Management

There are three core points from the discussions we have had thus far in this chapter, that are important to what you are about to read. First, for an organization to be truly resilient all five asset types must be evaluated to adequately determine asset resilience requirements. Second, as a factor of the type of assets under consideration, it would follow that some of the requirements can be satisfied by implementing security controls that involve additional security policy, while other requirements depend on enhancements to the configuration of assets themselves. Third, the resilience requirements that have been identified and prioritized affect the organization at the strategic level in addition to the ability to provide critical services that satisfy the established mission.

The point is, the organization cannot approach development and implementation of resilience solutions haphazardly. Whether the prioritized resilience requirements warrant the need for new controls or changes to existing control configurations must be made, care must be taken that previously defined asset configuration baselines are properly managed. Change management control must be in place that provides the oversight into who implements the solutions, when the solutions get implemented, and ensures that the service dependency among assets is considered. While software and technology assets can use the standard system life cycle approach to developing resilience solutions, satisfying requirements of the other asset types requires data driven approaches that encompass the participation of senior executives down to functional staff. Through our discussion of asset management in Chapter 2, we introduced you to the concept of CM. It is important to note that CM is always a two-part process—which you would basically view as "one shot setup." The first part relates to asset identification, labeling, and baselining during asset management activities. The second part, called configuration sustainment, involves managing the changes that take place to the established baselines. With that in mind, we will simplify our discussion here, building off the assumption that the first two process phases of CM are complete. A CM plan has been created, and configuration items (the integrated combination of baselined assets that provide the functionality of critical services and are managed through the CM process) have already been identified and labeled.

What Is Configuration Sustainment?

In defining configuration sustainability in the context of cyber resilience, care must be taken not to isolate the term for the maintenance of already existing asset configurations. A general definition of the word "sustainment" is "the act of supporting an organization through the process of maintenance." That definition was further put into context of the larger CM process back in the 1980s to characterize the change control of any form of IT activity. Today, the same practices used to provide control within the IT processes can be used to facilitate activities that foster cyber resilience within the organization. In short, through the process of requirements' prioritization we have identified changes that must be implemented in order to maintain consistency with the organization's resilience strategy. Those changes could likely change baseline configurations or create entirely new baselines. Through configuration sustainment, a plan is followed that provides the capabilities of managing updates to the baseline configuration items, in addition to providing the capabilities of recognizing new configuration items warranted by resilience requirements. Further, configuration sustainment provides the plan for assessing or testing the level of compliance with the established baseline configuration and mechanisms for reporting on the configuration status of items placed under CM.

Who Participates in Configuration Sustainment?

The three roles that are involved in configuration sustainment are those of the asset user, the asset custodian, and the asset owner. Recall that the asset owner is the entity that has purchased the asset or has ownership rights of the asset's user (this is generally the organization or one of the organization's supply chain partners). The custodian is the individual or group responsible for providing the functional capabilities of the asset to the user. For example, the custodian for a particular information asset could be the IT function. The user is that individual or group that utilizes the asset.

It would have been the custodian that, in the early stages of establishing a CM process, created a CM plan with the user and then ensured that it was understood, properly set up, and maintained at all levels of the organization. The custodian is also responsible for appointing configuration managers who have defined responsibilities for ensuring that configuration requirements are properly executed and maintained. Further, the custodian must also be able to assure product quality by performing occasional inspections of independent requirement solution implementation activities.

The user assigns a representative who has the proper authority to resolve all pending issues between his/her functional unit and the custodian. This representative has responsibility for approving proposals and concluding agreements with the custodian and for ensuring that his/her own functional unit observes all agreements.

Configuration sustainment incorporates the two processes of configuration control and verification control, which are implemented through three interdependent management activities and must be fit to the needs of resilience implementation process. The three activities are change process management (which is made up of change authorization, verification control, and release processing), baseline control (which is composed of change accounting and asset inventory management), and configuration verification (which includes status accounting to verify compliance with specifications).

What Are the Roles within Configuration Sustainment?

Each role in the process is assigned to an appropriate manager or management team. The configuration manager ensures that the requirements of change management are carried out. The configuration manager's general role is to process all change requests that have been generated through requirement prioritization processes similar to those we discussed earlier in this chapter, manage all change authorizations, and verify that the changes are complete. As we discussed in Chapter 2, the custodian also appoints a baseline manager who ensures that all configuration items in the asset CM plan are identified, accounted for, and maintained consistently with a specified identification scheme. That baseline manager also has responsibility within the context of configuration sustainment. They must establish a change management ledger (CML) for each controlled asset, record all changes, and maintain all records associated with a given asset. Important to note is that the CML contains some duplicate asset information to what is contained within the asset inventory. However, the ledger is necessary to provide a record-keeping mechanism for changes to the assets listed in the inventory.

The verification manager has the responsibility for ensuring that asset integrity is maintained during the change process. The general role of the verification manager is to confirm that items in the CML conform to the identification scheme established in earlier CM activities, verify that changes have been carried out, and conduct milestone reviews. The verification manager also maintains documentation of all reviews. The verification manager must guarantee that items maintained in the CML reflect the status of the asset at any point in time.

Control of Configuration Change

If organizations want to maintain resilient configurations for their assets, policies, and procedures in an environment where technology and business practices are continually evolving and the number and seriousness of threats is expanding, changes to configurations need to be managed and controlled.

Configuration change control is the documented process for managing and controlling changes to the configuration of all organizational assets, policies, and procedures that support the organization's resilience strategy. Such a process involves the systematic proposal, justification, implementation, test/evaluation, review, and nature of changes to the assets, policies, or procedures which include upgrades and modifications. Shown in Figure 4.10, the degree in which configuration change control is applied includes

- changes to any asset, policy, or procedure that affect the services being performed or the organization's ability to carry out its mission;
- changes to the configuration settings that may affect the performance of an asset;
- emergency/unscheduled changes;
- changes to correct flaws that limit the ability to provide resiliency.

Control must be enforced from the time the change is proposed to the implementation and testing of the change. Each step in the change process is clearly articulated along with the responsibilities and authorities of the roles involved.

The next several subsections describe a detailed perspective of the activities involved in controlling configuration changes and are discussed in the order in which they occur. Although, organizations always have the flexibility to determine which activities get performed, in what order, and at what level. It is important

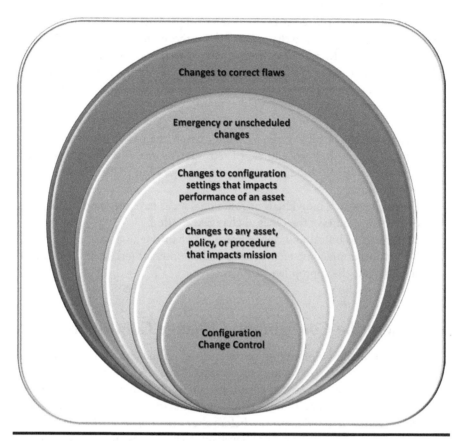

Figure 4.10   Configuration change control.

to mention at the outset of their introduction that the activities are normally implemented at the service level following predefined policies and procedures. Configuration change that affect the entire organization are implemented by the services affected in conglomeration, in order to meet the specific organizational needs. Nevertheless, completion of the activities associated with controlling configuration changes results in a planned approach to the implementation of access restrictions for change and documented configuration change control.

Implement Access Restrictions for Change

Access restrictions for change represent the enforcement side of CM. Configuration change control is a process for filtering changes for assets, policy, and procedure through a managed process. However, without access restrictions it would be impossible to prevent someone from implementing changes outside of the defined process. Access restrictions, therefore, provide an adequate means for enforcing configuration control processes by predetermining individuals or organizational functional areas with access to the assets or the authority to make changes to policy and procedures in order to make changes. Access restrictions for change may also include controlling access to additional change-related information such as change requests, records, correspondence, change test plans, or results. To implement access restrictions for change:

1. Determine the possible types of configuration changes that can be made to assets, policies, and procedures. In retrospect, some of this work may have already been done through vulnerability analysis and resilience requirement prioritization.
2. Determine which individuals or functional areas have the appropriate access or authority to make the types of changes required.
3. Implement technical and nontechnical mechanisms (e.g., role-based access, file/group permissions, etc.) to ensure that only authorized individuals are able to make the appropriate changes.

Implement the Configuration Change Control Process

A well-defined configuration change control process is fundamental to providing the necessary capabilities for achieving organizational cyber resilience. Configuration change control is the process for ensuring that configuration changes to assets, policies, and procedures are formally requested, evaluated for their security impact, tested for effectiveness, and approved before they are implemented. Although the process may have different steps and levels of precision depending on organizational resilience strategy or the amount of impact resilience has on service-level functionality, it generally consists of the following activities (Figure 4.11).

1. **Request the change**. A change request may originate from any number of sources including the owner, custodian, or user of an asset that is resilience dependent. Likewise, a proposed change may also originate from a supply chain partner or third-party vendor. There is no standard format for a change request. However, it will normally provide pertinent information such as
   - change description;
   - change justification;
   - urgency of change;
   - assets or policies requiring change;
   - expected resilience impact;
   - expected functional impact;
   - and funding requirements (if any).
2. **Record the request for the proposed change**. Once received, a change request is formally entered into the configuration change control process

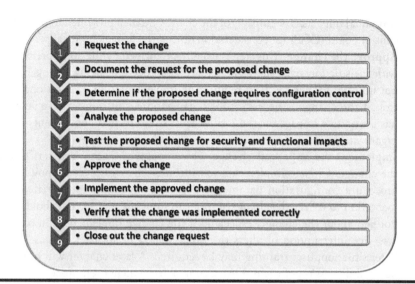

1. • Request the change
2. • Document the request for the proposed change
3. • Determine if the proposed change requires configuration control
4. • Analyze the proposed change
5. • Test the proposed change for security and functional impacts
6. • Approve the change
7. • Implement the approved change
8. • Verify that the change was implemented correctly
9. • Close out the change request

Figure 4.11   Configuration change control process.

based on predefined organizational procedures. Such procedures could use paper-based requests, emails, a help desk (for IT related assets), or some form of automated tools providing the capacity to track change requests, route them based on project planning processes, and allow for electronic acknowledgements/approvals by the appropriate organizational authorities.

3. **Determine if the proposed change requires configuration control**. It is not uncommon for some types of changes requiring minimal configuration change to be exempt from configuration change control all together or preapproval as defined in the CM plan. If the change is exempt or preapproved, that must be indicated on the change request so that the change can be made without further analysis or approval. However, appropriate plans, baseline configuration, policy documentation, or other records may require update of the nature of the exemption or preapproval.

4. **Analyze the proposed change**. This is one of the most important steps of the configuration change control process with respect to cyber resilience. Organizations spend significant resources developing and maintaining the resilient state of assets, policies, and procedures; failing to properly analyze a change for its security impact can undo this effort and expose the organization to risks that they were originally trying to avoid. In essence, this step provides the linkage between configuration change control and improved resilience. At a minimum, changes must be examined for impact on resilience and for mitigating controls that can be implemented to reduce any resulting vulnerability.

5. **Test the proposed change for security and functional impacts**. Testing confirms the impacts identified during analysis and can potentially reveal

additional impacts. The impacts of the change are then presented to the change control board (CCB) for consideration during the formal approval process.

6. **Approve the change**. This step is usually performed by the appropriate CCB with jurisdiction over the asset, policy, or procedure being changed. It is not unusual for the CCB to require the implementation of additional controls if the change is necessary for organizational mission achievement but has a negative impact on the resilience of other related components of the organization.

7. **Implement the approved change**. Once approved, authorized staff makes the change. Change implementation includes changes to asset, policy, and procedure configuration parameters as well as updating appropriate documentation to reflect the change(s). Prior to implementation, stakeholders are notified about the change, especially if the change implementation requires a service interruption or alters the functionality of an asset. In case of the latter situation, user training may be required. A later chapter will provide a detailed discussion of the resilience control implementation process.

8. **Verify that the change was implemented correctly**. As implementation completes vulnerability scans, post-implementation resilience, and functionality analysis, reassessment of affected resilience controls is performed. Configuration change control is not complete and a change request not closed until it has been confirmed that the change was deployed without issues. Although the initial analysis and testing may have found no impact from the change, an improperly implemented change can cause its own resilience issues.

9. **Close out the change request**. With completion of the above steps, the change request is closed out according to procedures defined in the CM plan.

If configuration change control procedures have been defined by the organization, the asset, policy, or procedure owner interprets the procedures in the context of the defined resilience requirements and refines the process to make it practical to perform. It is not uncommon for these changes to the process to need approval by the policy and procedure CCB according to established CM policy.

It is important that the staff who support and provide functional support of resilient assets be active participants in the configuration change control process and are aware of their responsibility for following it. If significant business process reengineering is needed, for example, updating operational procedures may be necessary simply to provide the resilience of a given asset.

Record and Archive

Once the change has been analyzed, approved, tested, implemented, and verified, the organization ensures that updates have been made to supporting documents such as technical specifications, policy manuals, procedure guides, and baseline configurations. Since the changes made relate to preservice organizational resilience, it is wise to use this opportunity to evaluate and update the content contained within security-related documentation such as system security plans, risk management plans, risk assessments, vulnerability analysis reports, and organization-wide facility, and IT long- and short-term project schedules and related milestones. When evaluating risk management protocols, configuration changes may level certain information, software, or technology into a higher risk category, in which case, a system reauthorization may be required.

As changes are made to baseline configurations, the new baseline becomes the current version, and the previous baseline is no longer valid but is retained for historical purposes. This is true of all assets, policies, and procedures. In case of assets, if there are issues with usage rollout, retention of previous versions allows for a rollback or restoration to a previous resilient and functional version of the baseline configuration. Additionally, archiving previous baseline configurations is useful for incident response and traceability support during formal audits. These topics will be discussed at length in a later chapter.

Resilience Configuration Management Monitoring

If the resilience required assets, policies, and procedures of an organization are inconsistent with approved configurations as defined by the organization's baseline configurations, or the organization's asset inventory is not up-to-date, the organization runs the risk of being ill-advised of potential vulnerabilities, and in turn may not take necessary corrective actions to otherwise limit those vulnerabilities and provide a necessary level of cyber resilience to protect it from possible threats. Monitoring activities provide the advantage of better visibility into the actual state of resilience at all levels of the organization, and also support wide-spread adoption of resilience CM policies and procedures. The activities associated with resilience CM monitoring also affords the organization appropriate input into the larger scope resilience continuous monitoring strategy.

It is the resilience CM plan that contains the details that the organization uses to implement the configuration monitoring strategy. The activities provided in that plan confirm that the existing configuration is identical to the current approved baseline configuration, that all assets in the asset inventory can be identified and are associated with the appropriate services, and, if possible, whether there are any unapproved configuration items. Unapproved configuration items (CIs) are generally a major threat to the underlying resilience of the organization. The reason is because they rarely have been updated, are not configured using the approved baseline configurations, and are not assessed or included in the authorization for use. For example, if a software engineer uses a specific software program used only for the purpose of testing and then forgets to remove it, or if an employee is provided access to a specific facility but does not have management consent, the organization may be vulnerable without being aware of it.

Assessment and Reporting

One of the most vital activities of resilience configuration monitoring is assessment and reporting. In cases where an organization has identified large number of assets, policies, and procedures that require implementation of resilience controls, the only practical and effective solution for monitoring activities is the use of automated tools that use standardized reporting methods. Organization-wide, there may be many requirements for resilience and in turn many baseline configurations. For the organization to manually collect information on the configuration of all assets, policies, and procedures and be able to assess them against approved baseline configurations is not an effective way to approach a resilience strategy, or even possible in most cases. Automated tools can provide the functionality the organization needs to facilitate reporting for resilience specific information and can extend to the capabilities of event management that management can assess or can alternatively be formatted into other reports on baseline configuration status. However, the organization must be careful to use appropriate mechanisms for collecting and analyzing the results generated by these automated tools in order to justify false positives.

There is not one specific approach for performing resilience CM monitoring. Rather, depending upon the requirements of resilience, monitoring may be performed in several ways such as:

- Using automated tools to scan for assets not recorded in the asset inventory. For example, after testing of a new vendor-supplied software product, the IT staff might forget to remove it from the system. If it is not properly configured for the system, it may provide access to the system for an attacker. A scan would identify this software product as not a part of the asset inventory, enabling the organization to take appropriate action for removal.
- Automated tools can also be used for scanning to identify inconsistencies between the approved baseline configuration and the actual configuration for an asset, policy, or procedure. For example, a security engineer might implement a security patch to an intrusion detection system (IDS) but forget to update the baseline configurations of that system or the software and hardware impacted by the new patch. A scan provided through an automated tool will likely identify a difference between the actual environment and the description in the baseline configuration enabling the organization to take action. Similarly, a new vendor-provided software application might have been installed on the workstations of specific staff within a given service area. It is not uncommon for application installations to change system configurations. In this case, the installation changed configurations exposing

information that previously had resilience-provided security controls applied on this, exposing them to attack. A scan would identify the change in the workstation configuration, allowing the appropriate individuals to take action.

■ Implementation of automated change monitoring tools (e.g., change/CM tools, application white listing tools). These tools are very popular in the IT industry where all development is done within a system environment that mirrors what the user has access to within production. Unauthorized changes to hardware or software within production, in this case, may be an indication that the systems are under attack or that resilience CM procedures are not being followed. Automated tools are available that sit between development and production in order to monitor system changes and alert appropriate staff if unauthorized changes occur.

■ Performing quality audits and review records in order to identify possible unauthorized changes that occur.

■ Performing integrity checks throughout the organization to verify that resilience baseline configurations have not been changed, either intentionally or inadvertently.

■ Reviewing configuration change control records to verify conformance with resilience CM policy and procedures.

The organization must be ambitious in their effort in ensuring that appropriate data is available that describes their assets, policies, and procedures such that the outputs from monitoring can be combined, correlated, analyzed, and reported in a consistent manner. Many automated tools provide a common language for describing vulnerabilities, misconfigurations, assets, and the policies and procedures that determine their use, as an obvious starting point for organizations seeking a consistent way of communicating across all services, regarding the underlying state of the organization's resilience status.

It is not uncommon for inconsistencies to be discovered as a result of monitoring activities. When that happens, the organization must take the appropriate corrective action. Such action may take the form of manual methods (usually in the instance of inconsistencies among policy and procedure) or through the use of automated tools. Examples of possible corrective actions include

■ implementing resilient security controls (e.g., taking from production unregistered software and hardware devices);

■ alerting appropriate staff of change details;

■ removing the change by reverting back to the asset, policy, or procedure from which the change was made;

■ updating the asset inventory to include newly identified assets and resilience requirements;

■ updating baseline configurations to represent new configurations.

As changes are detected and corrective actions implemented as a result of monitoring activities, those configuration changes must be documented. Specifically, the notations must contain the following:

- Who made the change?
- Was the change planned, or was it the result of an unexpected event that changed resilience requirements?
- Was the change previously detected and approved?
- Was the change done in a manner that is consistent with the CM plan?

Further, the organization must carefully analyze the results of the monitoring activities to gain better insight into why an unauthorized change occurred in the first place. There is a plethora of reasons why unauthorized changes can occur within the organization. Nevertheless, they all seem to stem from

- changes that occur accidentally or without knowledge that it has taken place until afterward;
- malicious cyberattacks and other disasters;
- individuals ignore the policies and procedures associated with configuration change control processes;
- individuals who have not been properly trained in the configuration change control process;
- failure to address errors that occur during the implementation of changes;
- a delay between introducing the change and updating the asset inventory and baseline configuration at the organizational level or for the affected service area.

By implementing policies within CM processes for analyzing unauthorized changes that have been identified through monitoring, the organization can not only more effectively identify resilience vulnerabilities but it can also get some valuable insight into any potential organization-wide issues regarding how the configuration change control process is being managed. If effective measures are in place for the organization to be made aware of such problems, the appropriate corrective actions can be made such as reengineering processes, implementing improved restrictions for change, or providing training on the proper procedures that must be followed throughout the resilience CM processes.

In addition to establishing policy and procedure for following the processes of resilience CM monitoring, appropriate policy and strategy must also be in place related to the outputs of the process. Such outputs must be made available to organizational management for the purpose of review and redefining the process as necessary so that the organization's resilience requirements continue to be met. Likewise, various types of reports are produced through the process that may be needed to support compliance with applicable federal laws, Executive Orders, directives, policies, regulations, standards, and other industry specific guidelines.

Chapter Summary

From the perspective of most practitioners that play a role in oversight and implementation of cyber resilience strategies, it comes as no surprise for them to admit that the elicitation of resilience requirements at the organizational and service levels (largely policies and procedures) together with the prioritization of asset types that must remain resilient in order for mission critical services to remain functional, is neither an art nor a science. It is often times a very tedious and politically charged process that requires those involved to have a collaborative and firm understanding of the organization's mission and objectives in combination with the services that achieve them. Those become the "critical services." Assets in the form of facilities, information, software, technology, and people support those services and therefore become a priority in determining and implementing the resilience requirements of all assets in the organization's asset inventory. The difficulty that comes about this process is fourfold. One, each asset type must be evaluated individually with regard to the resilience that type of asset must provide to the critical service. Two, there is not a one-to-one relationship between services and a specific asset. Often, services must be considered in combination with one or many others to determine how the asset supports each service and what must exist of resilience requirements for that asset in order to maintain the stability of each service. That is where organizations are likely to be challenged with the political wrangling between custodians and management within each service. Three, resilience requirements may not necessarily be achieved just with existing assets. Often the prioritization of asset resilience uncovers the necessity for the organization to invest in the obtainment of certain types of assets in order to fulfill resilience requirements. Four, the implementation of prioritized resilience requirements promotes change within the organization. A process must be in place to manage that change.

In Chapter 2, we introduced the concept of CM. There, we were concerned with identification and labeling of assets within the asset inventory. At the same time, the need for developing a configuration baseline for the assets was addressed. Once resilience requirements for each asset has been identified and prioritized, controls must be implemented to satisfy those requirements. Thus, changes are likely to occur to the baseline configurations originally established. The other side of CM is commonly called configuration sustainment (configuration change control, alternatively). Without proper procedures in place for managing the changes that take place to established configurations, the requirement for resilience of assets, policies, and procedures is compromised. Changes are often planned and intentional, but often change can occur unintentionally. Through configuration sustainability, monitoring mechanisms can be put into place in order to detect changes to resilient baseline configurations and take the appropriate actions to ensure that procedures were followed appropriately or that corrections were made to bring the baseline configuration back into alignment and ensure that proper change control protocol was followed going forward.

Keywords

Asset prioritization: The process of understanding the resilience requirements of each asset type and identifying the order in which those requirements are to be satisfied

Baseline configuration: The minimum characteristics that an asset, policy, or procedure must provide in order to satisfy the specification of resilience requirements

Configuration sustainment: The defined process that ensures that baseline configurations are met while managing the changes that take place to those configurations through change control, verification, and monitoring

Critical services: Those business services that must remain operational in order for the organization to continue achieving their mission and objectives

Mission assurance: The combined application of system engineering, risk management, quality, and management principles to achieve success of a design, development, testing, deployment, and operations process

Project portfolio management: An organization's grouping and management of projects as a portfolio of investments that contribute to the entire organization's success

Provisional impact level: The degree in which an information asset requires resilience controls in order to protect its CIA

Requirements elicitation: The process of gathering the necessary data useful in understanding the organization, its services, and the requirements for resilience necessary to maintain operability

Reference

BSI. (2014), *BS 65000, Guidance on Organizational Resilience*. Standard, British Standards Institute, London.

Chapter 5

Control Design and Deployment

Following this chapter, the reader will understand

1. the importance of organizational design in the creation and deployment of defenses;
2. the difference between security process design and classic system design;
3. the issues associated with the cyber resilience design and deployment process;
4. standard design steps to ensure cyber-resilient control architectures;
5. the general structure and intent of well-designed cyber-resilient control structures;
6. the large steps to implement a formal sustainment process for deployed controls.

Designing and Deploying an Effective Control Architecture

The design stage creates the requisite controls for ensuring the organization's priority assets. Therefore, the design phase satisfies the explicit control objectives for each critical asset. Formal design processes are the commonly accepted means to ensure the correct and proper deployment and sustainment of a control architecture. Control architectures underlie the overall strategic security governance process. Design creates an explicit infrastructure of substantive controls that are aimed at effectively satisfying the organization's stated mission, goals, and objectives.

Design prioritizes the business' security goals and then implements a full set of targeted control behaviors to achieve the highest priority objectives. Design analyzes and assesses the resultant control set to ensure that the subsequent infrastructure as a whole satisfies the critical purpose of cyber resilience. If the goal of verifiable cyber resilience is not achieved then the design process carries out the necessary data gathering and analysis exercises to understand why and then modifies the necessary controls or plug gaps.

The customary outcome of the design process is a formal and fully documented cyber-resilient architectural solution. Architecture is an industry term that describes a comprehensive set of fully interacting and conceptually discrete array of technical and managerial controls. A control architecture embodies specifically tailored technical activities and tailored business procedures. These comprise a coherent system. A formally documented architecture of such controls provides verifiable best practice assurance of the effectiveness of the security solution for a given collection of assets.

A control architecture establishes intentional security control over the organization's information assets. A well-designed control architecture ensures that a uniform set of demonstrably correct security behaviors are practiced at all levels of the organization. That comprehensive scope is required because information is a business-wide resource that affects every part of the success or failure of the organization, as an entity, not just the information technology function itself.

Thus, properly conducted, the cyber resilience design process is always under the sponsorship of executive management. Locating the process at the strategic policy level is critical because upper level managers are the only people who have the authority to ensure that the design will be properly researched, targeted, and embodies every requirement. It also ensures that all middle managers in the organization will support and ensure its adoption.

Unfortunately, cybersecurity responsibility is often limited to the information technology (IT) department. That is a short-sighted and dysfunctional approach to the protection of an organizational resource because the management of an information system operation rarely has anything to do with overall business strategy and it has no authority to enforce policies outside its domain. The inappropriate location of responsibility for the overall cyber resilience design process is one of the likeliest facilitators of failure, since massive change in attitude and behavior in the organization at large is necessary, in order to ensure the sustainability of a cyber-resilient solution.

Every architecture must be implemented to fit contextual requirements, shown in Figure 5.1. This necessitates a comprehensive process to understand the precise form of the threatscape as well as the specific life cycle requirements of the active system that the architecture represents. Therefore, the design process moves from threat and risk understanding through the design of active controls, right through to the implementation and testing of the resultant control array prior to its deployment and sustainment as a working system.

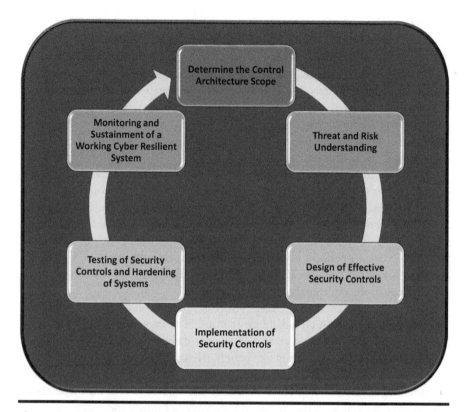

Figure 5.1 Design process of contextual requirements.

Setting the Protection Boundaries

The first step in that process is to define precisely what the control architecture will encompass and protect. As with any other type of security approach that process begins with establishing precisely what assets will fall within the general boundaries of the security scheme and what will be intentionally left out. It is much easier to create a comprehensive and stable architecture if the boundaries of the protection are well established and unchanging. However, that is rarely the case with intangible assets.

All forms of information are essentially value abstractions, even if they are written down, and the threats to all of them continually evolve. Moreover, the problem of designing substantive measures to protect abstract objects is aggravated by the fact that it is difficult to tell when a piece of information is being harmed. That is, information is the only asset that can be stolen or corrupted while the organization retains possession of the original item. So, it is hard to set proper boundaries with any certainty.

Unless the organization chooses to protect everything, which would be prohibitively expensive and operationally infeasible, boundary setting is based on the concept of prioritized perimeters. This idea was extensively discussed in Chapters 3 and 4. A simple protection perimeter secures the outer boundary of the space that is to be protected. This must be known before the design can be created. So, the first step in any form of boundary setting exercise is to explicitly delimit protected space.

Not surprisingly, the designation of a protection perimeter involves defining what the system will defend. Defining a proper perimeter is complicated by the feasibility factor. Feasibility is the likelihood that a task or purpose can be accomplished. With a control system, feasibility is judged based on whether the perimeter that has been demarcated to ensure all priority assets fit within the available resources and capabilities of the organization.

No matter how elegant the solution, the protection scheme is useless if it is not feasible. For example, sophisticated encryption schemes can be utilized to secure electronic messages. However, that requires that the message be important enough to justify the encryption, and that all participants have the capability to send and receive encrypted messages. Therefore, adopting an advanced encryption standard (AES) encryption solution for your daughter's email account might be a questionable use of resources, because it does not fit well with common sense as well as the practical constraints of teenage communication.

Moreover, that would be the case even if the encryption itself was correct and properly implemented. A circumstance that might change this would be if somebody deemed the information contained in your daughter's messages to be highly sensitive, like perhaps your daughter herself. That might not be an eventuality a parent would like to think about. But it might justify the added cost and trouble of employing biometrics to protect her communications. However, if there was no obvious reason to employ, the cost would probably not justify the means.

Consequently, a central part of the process of planning and deploying a practical cyber-resilient perimeter is the assessment and trade-off between known threats versus the financial and staff resources required to secure a given item. Five factors that might be considered in this appraisal would include defining, and are shown in Figure 5.2.

1. The exact level of criticality for each asset that could fall within a protection perimeter
2. The precise degree of reliable protection that is required for each asset
3. The likelihood and impact of all feasible threats within the threatscape
4. The accessibility and interdependency of each element within the protection scheme
5. The overall complexity and criticality of every asset with the protection scheme

Most of the decision is based on the value of the asset. The assumption underlying the creation of a cyber-resilient protection scheme is that it is perfectly permissible to leave nonessential or low-value assets outside the defensive perimeter. Nevertheless, if a decision is made to leave an asset out of the protection scheme that choice must

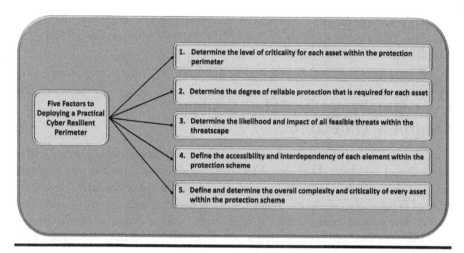

Figure 5.2   Five factors to deploying a practical cyber-resilient perimeter.

be based on reliable and comprehensive knowledge of the impact of its loss; this is an informed risk decision that is discussed extensively in the prior chapter 3.

A decision maker uses the asset prioritization process to maximize security resource use and to optimize the overall resilience of the protected space. It is vital to remember that the ownership of the information rests with the people who use it. These are called stakeholders. These people are critical elements of the design process because they are the ones who will be harmed if their information is damaged or compromised. Therefore, they are the ones to be consulted when the decisions are made about what assets are or are not included in the protection perimeter. For that reason, it is critical to include representatives from the organization as well as concerned third parties in the boundary definition process.

This consideration is particularly important in formulating a working cyber resilience architecture, because real-world threats and critical assets are dynamic entities. Even with precise knowledge of the form of the prioritized asset base, the borders of the protection perimeter are subject to ongoing refinement as threats and resources evolve.

The goal of protection perimeter setting is to ensure that all critical functions are adequately protected within the resources available to the organization. That requires an ongoing appraisal of the effectiveness of the protection with respect to resource commitments for the priority assets. Over time, the organization fine-tunes the deployment of information assurance controls. Proper modifications to the controls can be accomplished based on feedback from the operational units.

Therefore, the level of priority of a given asset must be assigned by the people who fully understand the consequences of its compromise. This rarely involves input from the people in the IT function itself. The working principle of the boundary setting process is that the business units are the asset owners. Therefore, these owners should provide the assessment of its worth. This axiom applies to functions and assets that are part of the business operation as well as other critical legal, contractual, and/or financial assets.

Conceptualizing the Cyber Resilience Control Architecture

A control specifies the precise collection of behaviors required to counter a credible threat to a given object. In practical terms, those objects are usually vital business assets or critical information. Because every organization has a different definition of what's vital and critical, control architectures have to be designed and sustained in the same way that every other tangible asset entity is constructed. That starts with the identification of protection goals, which is then decomposed down to a tailored set of controls.

The types of risks and controls and the methods for arranging them are detailed in Chapters 3 and 4. However, the controls themselves evolve out of explicitly understood policies which are established by a formal strategic planning process. The process for control design and deployment is universal and generic. That is, although the protection requirements for any given situation will vary, the elements of the process that is followed to develop the controls is always the same. Thus, all control assurance design and deployment processes embody five basic principles that should be very recognizable to anybody familiar with a system's work:

1. Top-down understanding and refinement
2. Progressive (or iterative) enhancement
3. Optimization based on feasibility
4. Continuous control
5. Measurement and assessment

It's a fact that the threatscape is varied. Threats to information assets will appear at various levels of day-to-day organizational functioning and in many forms. These threats range from complex electronic network exploits to the physical theft of items of value, which are perpetrated by people who knowingly, or unknowingly, take something important to the organization's resource base.

In large part, security control design and deployment are all about addressing issues raised by real or potential malicious actors. Their motivations may be individual, organizational, or even secondhand, so the practical importance of factoring aggressor actions into the control design cannot be underestimated. There are several methods available for estimating threats and impacts. These include:

■ *Misuse or Abuse Cases*—Constructing and using misuse cases can aid in understanding the concrete nature of how something can be compromised.

- *Attack Trees*—Attack trees set the attacker's goal at the top of the tree, and then explores branches that identify alternate or combined ways to achieve their goals. This provides a graphical way to portray potential attack paths which can then be countered by controls.
- *Physical Access Analysis*—If the attacker has physical access to the system then additional possibilities arise. Obviously, any logical source of physical compromise has to be identified and mitigated. Common cases of physical access include insider attacks and stolen devices.
- *Other Approaches*—This is the "other" category. There are always other previously unknown ways an organization can be exploited. That includes surreptitious entry of unknown attackers into work or equipment areas and subversion of personnel.

Utilization of these approaches should be considered when designing the control set for priority information protection. Consequently, the first step in formulating the control structure for a cyber-resilient response is to itemize and track all threats within the organization's technical or business threatscape, which might lead to the loss of any asset of any value. The outcome of this analysis process is an optimum control architecture. Five practical steps are involved in developing this architecture:

1. Organizational setup
2. Asset differentiation and valuation
3. Selection of the control set
4. Operational testing
5. Sustainment

Setting Up the Architectural Plan

The development of a substantive architecture of controls is a formal, strategic organizational planning activity. Therefore, it must be initiated and overseen in the same manner as any other business venture. First step in any project is the setup or launch.

The launch establishes the business need for a control architecture as well as its architectural characteristics. This is an essential part of the business operation. It is strategic and policy oriented. Because it will entail resource commitments and perhaps change organizational behavior, the purposes and value of a cyber-resilient architecture need to be identified and agreed on at the top by upper management. Then these aims must be publicized organization-wide and acknowledged by all affected stakeholders.

Such an extensive agreement process is essential because major cyber protection projects are both costly and disruptive. Therefore, the success of the overall project requires the total up-front buy in and commitment of the entire organization. That agreement could be documented in something as simple as a project charter that has been written under the auspices of an executive sponsor. Or, it might involve something as complicated as a detailed project specification, which supports the design and implementation of the control architecture.

The rule in exiting from this stage is that once an agreement has been reached, the direction that is captured in the agreement must be followed. Therefore, the formulation of an agreement to create a cyber-resilient architecture must NOT be taken lightly, since it will determine the form of the rest of the process.

Formulating the Protected Item List

Information assets are dynamic. So, they may exist in three different forms: electronic, paper, or human knowledge. That is, the same item could be recorded electronically, on paper, or in both mediums. Likewise, it can also be kept only in somebody's head and never actually recorded anywhere. If safeguarding the asset is that important then it might be necessary to take specific actions to secure it in all three of those states.

Before any protection scheme can be developed, the form of the asset must be known and categorized. We discussed that at length in Chapter 2. As we saw, the identification process involves a careful organization and classification of every item that will be placed under the control scheme. This is not a trivial exercise. It is prerequisite for subsequent control formulation because it establishes the "day one" state of the organization's entire array of potential items for protection. This was discussed in detail in Chapter 2, but because it is essential to establishing the overall control architecture it will be revisited here.

Terminology-wise, the aggregate set of explicitly classified corporate assets is termed a baseline. All the individual components that constitute that baseline are explicitly identified and labeled as part of the process of Asset Identification. An explicitly identified information asset baseline is an absolute prerequisite for the design and development of the control architecture, since those controls explicitly address the information asset requirements captured in the baseline architecture.

The control architecture is a concrete, formally documented organizational structure. The organization must design and deploy controls based on their priorities and their logical interrelationships. The resultant architectural design process is conducted as a top-down hierarchical decomposition of discrete behavioral elements that range in specificity from strategies, at the top of the organization right down to the actions of single individual control entities that regulate the behavior of a constituent target for protection, for a given strategy.

The decisions that establish the control architecture must be made by a range of participants. These can include the technical staff who have operational responsibility for the control target, all the way up to the executive stakeholders, who exercise ownership over that target's day-to-day organizational purpose. Thus, the process of control architecture formulation must be conducted in such a way that the subsequent control set functions systematically while not being influenced by external considerations such as revenue generation, which has no bearing on the effectiveness of the protection mission.

Each control that is placed in the hierarchy is given a unique and appropriate label, which is explicitly associated with its placement in the overall architecture of the control set. The label should be descriptive, and it must naturally relate the position of any labeled control within the overall architecture of controls to the baseline item of assets it is set to protect. Once established, the formal control architectural design is documented and maintained throughout the life cycle of the cyber resilience scheme.

Finally, because the information and its enabling technologies are continuously evolving, there must be a set of standard operational procedures set to manage even minor changes to the form of the control architecture. Once the control architecture is established, the organization must ensure that it continues to suitably address the threat environment. As we said in Chapters 2 and 3, ongoing risk assessments are critical to maintaining this alignment.

No control architecture operates on automatic pilot. Thus, regular and systematic risk assessments evaluate the potential for a given threat vector to adversely impact some aspect of the protection scheme. If the control architecture documentation is not continuously aligned with risk and updated in a timely, systematic, and disciplined fashion, it is likely that the substantive actions of the controls will become less and less effective, because their application will be less well-directed and applied to counter the behavior of a given threat.

Creating a Control Architecture from Best Practice Models

In most instances, control behaviors are based around and embody best practice. That is, the control is designed to operate in a way that is generally considered to be the best practical way of achieving a security purpose. These best practices are conventionally documented in the form of published expert models which are, in effect, a general specification of an architecture of controls that can serve as the basis for defining concrete security behaviors.

The well-defined and commonly accepted, all-purpose model architectures then serve as acceptable benchmarks to establish a substantive security response and then evaluate its *performance*. The "ideal" practices specified in the expert model provide the point of departure for substantive control formulation. That is, the organization's current operational goals and security objectives are compared to the benchmark ideal model to determine how close the organization currently comes to satisfying the requirements of best practice as defined in the model, as well as what needs to be done to improve its security status.

Any model of best practice can serve as the point of reference for creating a control architecture. The only rule is that the model must be commonly accepted and provide a reasonably accurate basis for judging whether the control architecture remains effective and relevant. Obviously, the impact of any gap between the current control practices and the stipulations of a best practice model should be considered in the light of whether they represent a potential security vulnerability. Once all deviations from the dictates of best practice have been identified and evaluated, decisions can be made about the specific operational mechanisms that will be deployed to safeguard the organization.

Making Asset Valuation Real

Asset valuation is important to control design and deployment because there is a direct correlation between the resources that will be deployed to establish a given level of control effectiveness. Operational factors, shown in Figure 5.3, that effect that consideration include answers to the following two questions: (1) What is the level of value of each given asset to be controlled? (2) What is the exact resource commitment that will be required to assure it?

As we said in Chapter 4, these questions need to be answered because there are never enough resources to secure everything of value in the organization.

Thus, the economic and operational value of each asset must be known in order to decide what priority to assign it. Since the design of the controls will be based

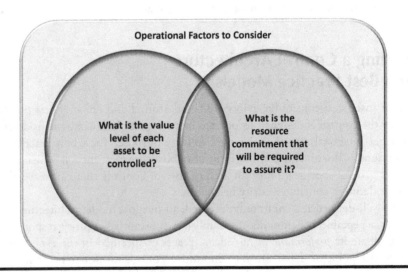

Figure 5.3 Operational factors to consider.

on the degree of rigor justified by the asset's inherent value, the actual process of designing and implementing the control can only be determined once its priority is known.

There are many ways to determine asset value. Most of these approaches are taken directly from the practices employed in standard economic value analysis, such as the Balanced Scorecard, Economic Value Added and Economic Value Sourced (EVA/EVS). Because these models are derived from the realm of business, the critical success factors must address and be relevant to the dictates of the business operation. The benefit of an economic valuation approach is that it helps the organization identify the precise scope and implementation sequence that is justifiable to ensure that the highest priority assets are ensured first. Criteria used to make that decision are based strictly on corporate strategies for value enhancement and preservation.

It should be evident that as data are collected and refined over time, the organization improves its effectiveness in doing these evaluations and therefore it increases its ability to protect corporate assets. The eventual outcome of this phase is an empirical understanding of the value trade-offs between all the major items in the organizational control architecture. This knowledge continues to regulate the boundaries of the system as well as the control implementation and sustainment priorities.

The Cyber Resilience Architectural Design Process

The common process of architectural design involves two separate but collaborative subprocesses. The goal of the first is to fully embody the outcomes of the prioritization phase into a single effective control framework. The eventual outcome of this phase documents a complete top-level control framework for a unique instance.

To do this successfully, the designer must produce an explicit and consistent across-the-board control architecture blueprint using the results of the Classification and Prioritization stages. That control framework will integrate all the basic components and the relationships involved into a single solution. The aim is to be able to reliably understand the relationship of every priority item in the organization through the mapping of analogous design element to its protection. This is a comprehensive, top-level view with the elements further explicated in the detailed design phase of the process.

Finally, the top-level architecture must be fully and completely documented in an organizationally standard design statement, which acts as a blueprint. That blueprint provides full visual understanding of the architectural details, components, and interrelationships simple enough that decision makers can see the "big picture" when it comes to comprehending the control structure. This is all defined and documented in a top-level design that shows each element, their external and internal interfaces, and their priority orders for implementation and sustainment.

As the control architecture evolves, the designers must also develop a means of validating and verifying the tangible outcome of the process. That also includes the means of assuring that all control elements have been adequately integrated into a single solution. This implies that all the components and their interfaces must be evaluated based on a standard set of criteria that document proper support and traceability to security goals, external and internal consistencies of the integration, appropriateness of the sustainment process, and feasibility, when it comes to meeting the overall goal of cyber resilience.

Designing Substantive Control Measures

Each of the components in the architecture must be assured by substantive control measures. This implies that the risk mitigation actions must be turned into a planned and intentional set of standard operating procedures. Those procedures ensure that each identified risk is addressed by a robust set of control behaviors. Therefore, the first step in that process is the visualization of specific risks and their concomitant attack vectors for each of the priority components. This is then followed by the deployment of a substantive control that is aimed at addressing each of those parts.

The first step in the control development process involves establishing the specification of behaviors required. The requirements are in large part derived from the dictates of the control system's environment and that system's interactions with environmental concerns in order to achieve a certain set of desired results while avoiding others. Control designers pay special attention to threats that have been identified as affecting priority elements of the organization.

Security requirements specify all the properties required to ensure the protection of a high-value organizational asset. In general, what this means is that these requirements enforce criteria related to availability; integrity including authentication; confidentiality including anonymity, privacy, and intellectual property rights; and accountability including audit logs and evidence of non-repudiation.

Resilience requirements identify not only what must or should be executed by a given control or set of controls but must also identify what should be tested for to obtain sufficient evidence to support that assurance. That includes proof that the external behavior of the control system meets the constraints of organizational security policies. In general, these include:

Traceability—e.g., the ability to trace protection requirements, required features, system security policies, and other types of control requirements to the action specified by the design for implementation. Security control cannot exist in isolation, so traceability must also be upward and outward to any environmental requirements in the threatscape or the documented outcomes of the classification and prioritization phases. That includes the goals, means, and processes for estimating potential for damage to assets that are encompassed within the security perimeter.

Assets must be protected in many situations across their useful lives, and they may exist in a number of different places, each of which could imply a different set of security control requirements including ones for

■ security of the information assets;
■ protection of tangible information processing assets;
■ communication security;
■ redundancy, backup, archives, and recovery records, including those for continuity;
■ system logs;
■ mobile computers and devices.

The assets to be protected usually imply explicit needs for

1. data protection;
2. protection of software-related artifacts and supporting data throughout their life cycle;
3. human and organization protection for users and operators;
4. security of physical and computing assets in the operating and user environment.

Ideally, the control set should address relevant threat entities or attackers for any assets deemed worthy of priority protection. As we have seen, the security protection needs are based on the organization's stakeholders' valuation of the assets and the assessment of the consequences of successful exploits.

The motivations and available resources of threat entities themselves must also be established. This includes both current and future capabilities of potential attackers. The rate and degree by which every feasible threat entity will improve its capability is also something that must be considered. That is because the defensive measures must directly counter attacker capabilities now and in the future. Therefore, part of the design function boils down to identifying or postulating attacker capabilities. This identification will help forecast how the control set needs to be evolved in terms of future robustness. The ability to specify robustness should evolve with the design. Thus, security design requirements continue to be created during design activities.

Finally, usability is an important aspect of control design, not only because it reduces user mistakes but also because it ensures efficient and effective acceptance of security features. Security is likely to be systematically bypassed by users if it places too much perceived impediments on a user intent on accomplishing their tasks.

Process and task analyses are the starting points for understanding the precise form of the requisite real-world control set. Unlike other forms of technical design, there is no commonly accepted methodology for understanding and abstractly depicting a cyber-resilient control architecture. Therefore, it is suggested that the best means of documenting the form of the control array is by representing the architecture using multiple process viewpoints.

Examples of potential viewpoints include the *functional view*, e.g., the behaviors the architecture will exhibit; the *integration view*, e.g., the interrelationship and interdependencies between the various assigned controls; the *physical view*, e.g., the spatial/timing representation of the various control actions; and the *development view*, e.g., the representation of the control architecture for the purpose of implementation, as shown in Figure 5.4.

The premise is that this intersecting set of viewpoints willfully and sufficiently describe the shape of the control architecture, sufficient to underwrite its practical implementation. The aim is to present the architecture in enough detail to support the development of a substantive set of organizationally standard descriptions of the design and implementation features, which can then be kept and used for the implementation and future evolution of the design. Shown in Figure 5.5, there are twelve specific viewpoints that might apply.

1. *Context*: describes the services provided by each control from one specific perspective. The perspective represents the way that each individual external actor or stakeholder will utilize or approach the asset.
2. *Composition*: describes the way each control will interact with every other control. This is supported by functional decomposition of the overall control problem into its necessary components.

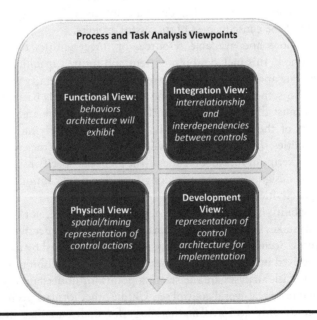

Figure 5.4 Process and task analysis viewpoints.

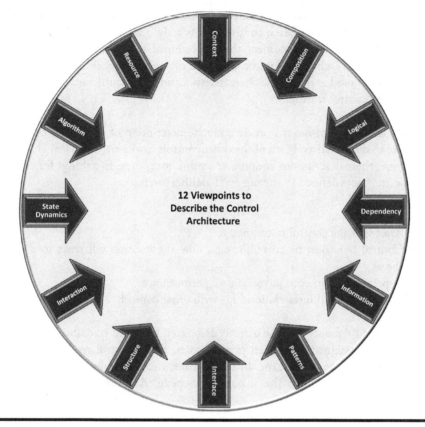

Figure 5.5 Twelve viewpoints to describe the control architecture.

3. *Logical*: describes the planned or intended processing or functioning of the control within the control array, given its placement.
4. *Dependency*: describes dependent relationships between the control and control set, in terms of their functional association in the array.
5. *Information*: describes the necessary information requirements for control operation, including stored and throughput data.
6. *Patterns*: describes the overall conceptual architecture and any influences from contextual sources or reuse.
7. *Interface*: provides the logical relationship between a given design subject and all other objects in the design. This description includes the details of external and internal interfaces for each design subject.
8. *Structure*: provides a description of the overall internal control functions and organization of the control array.
9. *Interaction*: describes the summative relationship between control items, e.g., how does each control contribute to the overall outcome of assurance.

10. *State Dynamics*: used to document the modes, states, transitions, and responsiveness to actions of each control and its activities.
11. *Algorithm*: used to document the programmed functioning, internal details and logic of each control in the array, and the operation of the array as a whole.
12. *Resource*: used to model the characteristics and specific utilization of resources by the control feature.

Detailed design descriptions provide a precise description of what each control is supposed to do as well as which of the other controls it will interact with. Shown in Figure 5.6, these descriptions contain, at a minimum, these five things for each of the priority assets defined in the top-level architecture:

1. Control name
2. Asset that the control is assigned to
3. Control behavior, by stimulus, e.g., how the control will react to a given threat
4. Expected outcomes to judge control performance
5. Interactions and interrelationships with other controls

These five requirements must be directly associated with and traceable to a control that has been assigned to a given priority asset. Since control requirements will no doubt vary as the threatscape shifts over time, the design descriptions must be revisited on a regular basis. This is normally scheduled as part of the sustainment planning process.

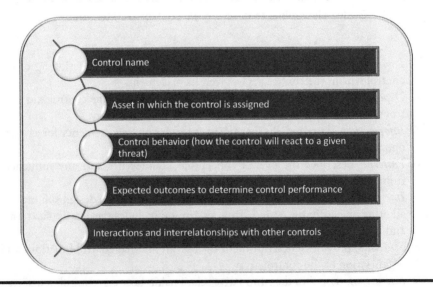

Figure 5.6 Control descriptions for each priority asset.

In the architectural design phase, the organization also produces a detailed set of testing and review qualification requirements for each priority component in the design. And, once these are known, the schedule for testing each component is laid out. Ultimately, all of the deliverables in this phase can be evaluated based on the representative criteria of traceability, external and internal consistencies, appropriateness of the methodology, and feasibility. Following the successful completion of testing, the organization produces an Integration plan. That plan is then applied during the architecture phase.

The Cyber Resilience Detailed Design Process

The detailed design of the controls is always unique to the situation. It is based on the operational purposes of a given protection target as well as the contextual threat picture that the target is embedded in. That can vary from place to place for the same type of target. Therefore, it cannot be stressed enough that detailed design of the control set must be approached on an individual piece-based, supported by the full understanding of the organization's real-world assets, risks and priorities. Nonetheless, there are certain principles that nearly all situations have in common and those can be generally applied as means of ensuring a consistent design solution. These underlying principles include:

- *Abstraction*: a means of breaking the protection requirements down into smaller related parts in order to document in detail the requirement at its simplest and most fundamental level.
- *Cohesion*: establishes a consistent relationship between the actions to be taken or procedures to be followed, traceable to a desired outcome for a single control.
- *Coupling*: establishes relationships between the various controls that are aimed at achieving a given protection purpose. This involves multiple controls by definition.
- *Decomposition*: the practice of addressing a specific problem situation by breaking the actions required to address it into smaller consistent parts to achieve a logical result.
- *Encapsulation*: the summation into a larger stated design outcome the related protection aims for a given control.

Designing the Individual Controls

The individual controls that will populate the cyber-resilient control architecture are deliberately designed entities. Whether the control is a policy, a specific recommended action, or an electronic operational feature, each of them is designed to specifically address an identified risk or threat. Thus, control design occurs once the threat picture is established.

The controls themselves are behavioral in that they specify a specific action to be taken or programmed response that will occur, given the fulfillment of certain criteria. These actions or responses are preprogrammed, and must happen as intended by the design.

That requires an item-by-item evaluation of each information element and its attendant threats, to define and formalize the requisite actions to be taken. That assessment evaluates every possible threat vector and then deploys a set of practical actions to ensure that no undesirable contingencies occur. This design and deployment planning is done for every information element in the asset base.

The actual specification, design, scheduling, and installation of a working set of control behaviors is a very intricate and involved process. Therefore, it comprises most of the actual day-to-day work that is involved in establishing a substantive cyber-resilient defense. The outcome of that process is the creation of an infrastructure of explicitly targeted security control behaviors for the organization.

Because it is such an important element of establishing and sustaining a cyber-resilient architecture, we are going to devote most of the rest of this chapter to describing the process for infrastructure creation and sustainment. However, before we do that it is important to emphasize that each essential security control behavior must be engineered for optimum performance. Then the collection of those controls must be maintained in provably harmonious agreement with the security goals of the organization.

The individual controls embody and enforce the organization's current aims when it comes to asset assurance. Each control is documented and kept under strict organizational oversight. This oversight is enforced in the exact same way that every item of value is accounted for in an organization.

The design and the associated control outcomes are maintained directly traceable to each other. Every protection target can be reliably referenced to its attendant control set. Maintaining trackability is the only way to be certain that the control behaviors satisfy strategic security requirements. Therefore, the maintenance of explicit tracking between control operation and security intent is an important overall task of the cyber resilience control design function.

Operational Testing

The need to assure the validity and correct functioning of the control array is a critical requirement for the cyber resilience function. Once the control set has been designed and implemented, it must be proven effective. This is a necessary condition to ensure that the proposed controls meet the detailed requirements for safety and security of operation as well as all performance criteria established in the strategic plan for cyber resilience. Operational testing takes place after the controls have been designed and deployed within the operational environment. The process itself is called "operational testing" because it is assumed that the control architecture is already functioning in that environment.

From an assurance management standpoint, operational testing is a beta test process. It entails the ongoing analysis of actual outcomes against the performance criteria that were established in the conceptual design phase. Operational testing is planned, implemented, and administered just like any other testing activity. It utilizes the assumptions embodied in the conceptual design of the control architecture as the basis for doing the assessment; however, unanticipated operational issues that are identified after the actual implementation of the control set may be added at this point. The aim is to be able to say with assurance that the architecture and its aggregate set of controls are operating effectively given the security goals of the organization.

Operational testing is a formally scheduled and executed assessment process. It requires establishing a specified period for conducting tests as well as a reporting and decision-making mechanism that is specifically set to respond to the outcomes that are obtained from the testing.

Because the overall goal of this step is to ensure the correctness of the final product, each control must be tested against a set of assigned criteria that have been designed to assess, without any ambiguity, control performance. This level of specification ensures a systematic evaluation of each critical component in the cyber resilience scheme. As such, each control must have a behaviorally observable result associated with it.

Finalization of the Control Architecture

Once the testing is complete, the aggregate results are analyzed in order for the organization to be able to sign off on the control architecture. The organization employs the outcome from the operational testing phase to understand and authorize the operational deployment of the control architecture. The authorization is based on the observable performance of each control against established criteria, as well as any assumptions about resource commitments and cost that were part of the design process.

The operational controls constitute the everyday form of the cyber resilience architecture. The system can be considered to be mature at this point. Therefore, the formal representation of that architecture is maintained under strict configuration management control. The operational system that emerges at this stage constitutes the organization's specific cyber resilience utility. That system is maintained as a functional organizational resource, and it is in effect the cyber resilience approach.

A Footnote: Handling Exceptions

All types of formal security control operate under the dark star of process entropy. Process entropy happens when intense competitive pressure is combined with the mind-numbing pace of technological change. These two factors tend to shred discipline, and as a result cause well-defined control processes to inevitably fail.

Process entropy is inevitable where unanticipated risks or changes to the threat environment are a practical given. Unforeseen or escalating attacks, changes in organizational behavior that invite risk, or alterations in the attack surface of the system because of new technology are all examples of this. Consequently, a proper system of cyber resilience controls has to include a well-defined mechanism for making a disciplined response to all forms of unplanned or unanticipated threat.

A systematic problem-solving approach is the proper response to such threats. This is generally termed an "exception handling" or an "incident response" process. These processes are called out once a new or unexpected threat is identified. In practice, this type of standard operating procedure is formalized as part of creating the cyber resilience architecture.

Its procedures are invoked when the organization encounters a problem that has not been planned for. Nevertheless, even though such problems might not have been thought about, or encountered before, it does not mean that effective controls can't be brought to bear on the problem in an effective and efficient manner.

A detailed and formal process to handle unplanned for exceptions is important in today's environment because of the increasing sophistication of attacks. Since attacks are becoming more complex, the organization's control architecture has to be able to be adjusted and respond to novel methods of attack. That response should embody the following attributes:

- Timely—response times ensure effective remediation
- Responsive—the response is evolved directly from the threat
- Disciplined—the process is structured and followed systematically
- Usable—the process involves all types of users in the solution

Timeliness is valuable because by definition, novel attacks or the exceptions to control planning are unanticipated. This is one of the reasons we have concluded that defenses built around perimeter control technologies that rely on preestablished patterns and rules are ineffective.

Responsiveness, discipline, and usability are functions of practice. Responsiveness is an important quality because it is necessary to ensure that the defense reacts to the threat in the optimum manner. *Discipline* is the execution of intentional actions based on well-thought-out, standard procedure. *Usability* is important because the effectiveness of the defense is typically proportional to the number of people who are involved. If incidents can only be recognized and responded to by a few people, it is likely that something new or unique will slip past them. If, however, each member of the user community is knowledgeable, aware, and involved, it will be much more likely that unplanned or unanticipated events will be immediately identified.

Communicating Organizational and Technical Direction

Finally, it should be clear at this point that the essential condition for the successful creation and sustainment of an effective cyber resilience control architecture is for all members of the organization to have common understanding and buy in for the behaviors that are expected of them. There is no way that the organization can hold an individual accountable if he or she was not given adequate information on how to behave. Therefore, to some extent, the success of the system centers on establishing and maintaining effective communication. All participants must understand the rules of behavior, because cyber resilience control schemes are complex and subject to interpretation by the participants. Participant understanding must be controlled to some extent to ensure consistent outcomes.

Following the operational testing and assurance phase, the practical control array must be formally implemented or embedded in the organization. This is a particularly critical and frequently overlooked planning requirement for the control deployment process. That is the case, because the control set is still essentially abstract in that it is not operating within its intended context. Thus, the choices made in this part of the process must be based on a detailed understanding of the local size, scope, and complexity of the day-to-day operation.

The principle here is that although there might be a logical set of activities that might be performed, it does not mean that the organization should undertake all those actions in their entirety for a given project. In fact, at the local level, the organization needs to step back and decide about the appropriate processes and procedures, methods and tools, and the resultant end technologies that are appropriate and practical for a given setting.

Following analysis of the local situation, the deployment process tailors the control architecture to fit the needs of a given assurance situation. That implies the need for tactical planning within the overall policy and procedure framework of the deployment project. Here, the people who are responsible for the actual focused deployment of the practical control set will formulate, document, and execute a plan that will be sufficient to assure the proper performance of all the essential activities specified in the control set activity and task requirements.

This plan must be specific, in the sense that it must describe any standards, methods, tools, actions, and accountabilities associated with the maintenance of the safety and security of a given protection target. These plans can be expressed at any one of several levels of detail. However, there is always at least one overall, practical operational plan for the entire cyber resilience control architecture.

In addition to that plan, there may be focused plans aimed at mapping a day-to-day control activity to an itemized protection requirement. The scope of those plans might be as modest as the specification of a given set of repetitive tasks to be performed on a routine basis. Or, it might describe an overall policy to guide decision makers on a given protection issue, such as access control.

However, at whatever level the specifications of behavior are defined, a comprehensive plan that covers all conventional eventualities is essential. That is because the plan will provide a mutually understood, organization-wide basis for evaluating process performance as well as ensuring knowledgeable management control. The benefits of a fully detailed and documented plan have been well supported by numerous studies over time.

Implementing and Sustaining the Solution

The purpose of the control design process is to develop and maintain an explicit model of the control architecture that includes precise specification of the control architecture's required behavior. This specification is then used to operate and sustain the practical control system. In many respects, control design is the single point where the organization can intentionally ensure a high level of security and integrity. That is because control design requires consideration of all the potential threat and resource considerations and the creation of a real-world organizational response to each of those threats.

The overall solution typically depends on the identification, deployment, and sustainment of standard security controls, which have been assured for correctness. The standardization of how these controls are produced ensures that the behaviors they exhibit embody common properties of quality and security. In order to ensure that correctness, the designer selects and utilizes a standard domain model and domain architecture. The control domain itself is then defined in terms of its scope, boundaries, and relationships to other domains.

When conducted as a rational process, this is called control engineering. In control engineering, an explicit control model is developed for each particular security domain. That model captures the essential common and dissimilar features, capabilities, concepts, and functions of a given domain architecture and controls, including their commonalities and differences. Then the controls for each asset are assigned based on these characterizations.

Logically, the first step in the control engineering process is to create and execute a plan to secure the domain as a whole. To operationalize that plan, the control designer selects and formalizes the standard form of representation that will be used to communicate the control array. The standard form of representation will be used to capture and portray the overall domain architecture and its components. Generally, the form of representation is a commonly accepted modeling approach like unified modeling language (UML) or feature modeling.

In addition, these models have to be usable for the actual stakeholders, so as part of the process of deciding on a form of representation the domain engineer also sets up a formal procedure for evolving the model based on input from managers.

Control engineering is essentially an abstract design process, just like all other forms of engineering. Therefore, the first step in the process involves a definition of the conceptual boundaries of the domain and the relationships between that domain and all other domains. Then the control engineer identifies the current and anticipated needs of stakeholders within the domain. This is a practical step in that it helps delimit the protection targets that might fall within the control domain.

Once the boundaries and contents of the domain are understood, the domain engineer designs and documents the abstract model of the requisite controls. This is done using the representation forms selected in the first step. Along with the abstract models for the domain, the engineer also formalizes a vocabulary to be used to describe salient domain concepts. Then the domain engineer begins to classify, design, and document the explicit controls for the elements of that particular domain.

Once these models are fully documented, the control engineer evaluates the design along with the common vocabulary in order to confirm that they both comply with the necessity for consistent representation. In conjunction with this, the control domain engineer also carries out an end-to-end review with all relevant stakeholders. That review includes high-level managers, asset managers, domain experts, and users. Once the review is complete and the results are accepted, the control domain engineer passes the domain models along to the implementers for operationalization in the domain setting.

The representation is conveyed in the form of a control domain architecture. That architecture must be consistent with, and capture the design and operate in accordance with the organization's overall strategic goals. The domain architecture documents an explicit set of requisite behaviors for each control that has been designated for implementation. That specification is evaluated in accordance with the organization's overall security strategies. Following acceptance by the relevant managers, the control engineer undertakes a design review with all relevant stakeholders and asset managers who operate within that domain and submits the resulting finalized domain architecture to the appropriate parties for implementation.

Having turned the behaviors, which were specified as required by the detailed architecture, into a logical representation, the organization now needs to develop ways to make these specifications into practical everyday tasks. That is the role of the implementation plan. That plan must specify the means that will be used to integrate all requisite control behaviors into a single seamless day-to-day process. That includes all the everyday procedures and practices, information requirements, responsibilities and accountabilities and schedules.

The organization installs and tests each control behavior in accordance with a specified implementation plan and ensures that each installation satisfies the overall requirements of the cyber resilience strategy. In that respect, the implementer must develop and document a set of tests, test cases, including inputs, outputs, test criteria, and test procedures to assure the practical effectiveness of each installed control. Once an acceptable plan for installing and testing control behavior has been prepared, the implementer updates the relevant documentation as necessary.

The actual operation of any form of overall deployment is cyclical in that there is a distinct installation phase, an operational assurance phase, and hopefully a long period of formally sustained use. However, these are not naturally occurring events, they must be prearranged within a specified life cycle. Therefore, it is necessary for the organization to specify a length of time for each phase to occur. This specification comes in the form of a schedule that is prepared prior to the implementation of the system. There are several considerations that need to be captured in an effective implementation plan. These include a precise stipulation of

1. the exact set of activities and tasks that must be completed;
2. the sequence, timing, and requisite intervals for each activity and task;
3. the people who will be responsible for the completion of each activity and task.

Another important aspect of implementation strategy is the determination of how problems that might be encountered during the actual implementation of the control set will be documented and resolved. At the end of the planned installation phase, the behaviors of each individual control have been assured reliable, and they have been successfully integrated into a single effective response.

This is all done by plan. That plan should include, as applicable, a description of the control context/environment, a rationale for the planned behavior based on that description, plus all information and analysis procedures to ensure the effectiveness of control performance. The plan should identify the larger threat protection requirements that have been addressed by each control and ensure that all requisite control requirements are assured by some form of qualification testing. The plan should also determine the appropriate timing and sequence to begin installation and provide details regarding all testing and assurance procedures, staff roles and responsibilities, and the documentation that must be prepared and delivered to ensure hand over to the operational use of the system.

That includes all specifications of organizational test strategies and policies, the management of the overall control process including the design of larger organizational as well as local testing and review strategies. It also requires plans for monitoring and controlling the long-term assurance process. As the control implementation and deployment phase concludes, the closeout should include some form of statement that the control set is correct and the assurance mission will function as expected.

Sustaining the Assurance Infrastructure

The maintenance of a secure control architecture involves the development, deployment, and continuous maintenance of the most appropriate and security response to threats and changes. To accomplish this, standard security and control processes must be planned, designed, administered, and maintained. The aim is to ensure that effective leadership vision, expertise, and the correct technology solutions are available to assure effective security and control of the cyber resilience control architecture.

Standard procedures must be in place to continuously assess, integrate, and optimize that security architecture, and the security architecture must be maintained consistent over time. Therefore, future control assurance trends also have to be evaluated to define the long-term architectural strategy. Emerging threats and security trends also have to be assessed as they might impact the evolution of the security architecture, and resources have to be developed and made available to ensure the long-term effectiveness of the control scheme.

Since the human factor is critical, training and education must also be provided to lay practitioners to ensure a minimum acceptable awareness of the implications and requirements of the security control architecture. Moreover, cross-organization collaboration must be established to communicate security practices by the systematic development and implementation of training materials. These materials must be organizationally standard. Their coordination must be carried out by a formal process.

To ensure secure practice enterprise-wide, the most appropriate and capable set of policies, processes, and methodologies must be developed, deployed, and sustained. Strategic directions also have to be developed and evaluated as they relate to the acquisition of security architecture components. From this evaluation, rational decisions can be made on the most effective purchase of products and services and the in-house development of tools. The aim is to maintain a dynamically effective security and controls architecture. Thus, the most current security methodologies, processes, and associated documentation have to be researched, developed, deployed, and kept in an organizationally standard and accessible repository.

Finally, all this hinge on effective assessment. Therefore, metrics to improve methodology and process efficiency, usage, and results must be defined and analyzed. Security metrics must be standard. Modeling of application and infrastructure must be done from a security perspective using these metrics. There must also be metrics-based causal analysis to optimize the ongoing security process. Formal teams must be established and coached in how to apply the organizationally standard methodologies and processes. This item also assists with assessing compliance.

Ongoing security and control certifications of applications and infrastructure must be enforced. Security and control awareness, knowledge of policies, procedures, tools, and standards must be championed and promoted. Altogether, to minimize risks the infrastructure needs careful attention on its security aspects.

Operational Sustainment

The operational sustainment function is the payoff for all the design and development work in the preceding stages. Even though it is an everyday process, this phase should be conducted in an equally rigorous manner as all the other preceding phases, in the sense that the operational functioning of each of the requirements must be continuously evaluated for correctness and compliance with organizational goals and intents.

The long-term assurance of the control architecture is specifically defined and formally implemented through a Sustainment plan. Commitment to this plan must be rigorously maintained throughout the life cycle. At the minimum, this plan should specify the way changes to the control baselines will be managed.

The purpose of the sustainment function is to establish and maintain the integrity of the architecture of controls that has been established for that particular organizational asset. Control sustainment does this by developing a coherent strategic oversight and change management process. The specific aim of this process is to ensure the completeness, consistency, and correctness of each of the control elements in the formal control architecture. Consequently, once the control architecture has been established, the sustainment function controls the modification and integration of every baseline item. Sustainment also records and reports on the status of each of these items. And, it maintains a record of the monitoring and oversight functions.

Sustainment ensures the integrity of control baselines throughout their functional life cycle. It does this by means of an operational sustainment process. The sustainment process is a critical organizational undertaking, because it ensures the prior time and investment in cyber resilience that has been made by the organization. Because the sustainment period can extend for years, a strategic plan is required to guide and adjust the function over time.

The plan describes the organizationally standard sustainment activities and the procedures and schedule that will be followed to perform those activities. It also designates those roles and people who will be responsible for performing each activity, and the role and responsibility relationship with other individuals and entities. The plan enumerates, in unambiguous terms, the typical activities that constitute the formal sustainment process for that organization. That includes the specification of the procedures for performing any control modification activity that might involve interaction with another organization. The planning can apply external units, for instance, suppliers or subcontractors, because those entities might influence or participate in some aspect of the overall organizational security function. The end product of the planning stage is a complete and correct, fully documented life cycle-oriented, strategic direction for the performance of the organization's sustainment function.

As in all the prior sections, the organization is required to evaluate control performance on some type of scheduled basis using well-documented criteria for traceability, external and internal consistencies, appropriateness of operation, and feasibility. To support this process, it is necessary to conduct scheduled reviews in accordance with formal reviewing best practice. In addition, the organization should also conduct and document periodic audits in accordance with the general dictates of due diligence. Shown in Figure 5.7, the general outcome of any form of in-process evaluation of control performance is to answer two important questions:

1. Does control performance meet the stated criteria for proper execution as specified in the design documentation? The process of answering this question is known as verification.
2. Does control performance achieve the intended behavioral benchmark criteria as specified in the design documentation? The process of answering this question is known as validation or testing.

The primary purpose of periodic control testing and reviews is to provide reasonable assurance that the control array as a whole continues to function as planned. The inspection and evaluation can also target any optional areas of inquiry where performance under normal and abnormal conditions needs to be evaluated. For instance, this would take place when a new credible threat has been identified or a novel technology has been adopted.

The reviews and tests should be designed to evaluate compliance with the stated criteria for performance in the design documentation. But, they should also provide tracking and feedback information for management to ensure that their stated mission, goals, and areas of the protection scheme are appropriate. The scope and

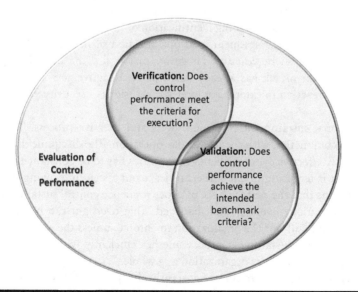

Figure 5.7   Evaluation of control performance.

detail of the tests and reviews must be tailored to the design and complexity of the assurance required. Obviously, the assurance needs of a top-secret agency will be different than those of a conventional business. But the requirement of continuous assurance of control performance for both these organizations remains. The data generated by tests and reviews is intended as an oversight and quality assurance support for the overall cyber resilience mission.

Sustaining the Control Architecture over Time

Cyber resilience is a state, not an end in itself. Therefore, it has to be maintained as such. Ideally, the cyber resilience process functions in the background of day-to-day operations, identifying and mitigating harmful or undesirable events as they appear in the threat horizon. This all takes place during the normal conduct of business. The aim of the long-term sustainment process is to ensure that the control architecture continues to be appropriate to the threat environment.

Because that environment is in a constant state of flux, the form of the controls in the control set as well as their overall applications and scopes will alter as changes appear and disappear within the threatscape. Consequently, along with the development of the control architecture, there is a necessity to plan and maintain a disciplined and systematic procedure to guarantee that the cyber resilience controls are consistently functioning as required. That is called "operational sustainment."

The overall function is conducted continuously in the background within the operating environment. That is the reason why sustainment monitoring is generally termed Operational Sustainment. Operational sustainment is a proactive process. Activities include ongoing identification of threats and vulnerabilities, and the attendant creation, assessment, and optimization of new controls as needed. Sustainment can also be reactive. However, that is not the desired state, since it implies that the attack has already taken place. Reactive activities include the detection and reaction to external or internal intrusions or security violations after the fact.

Operational sustainment is a continuous and iterative process. The process interacts in conjunction with the day-to-day operation. The discipline that is necessary to ensure effective cyber-resilient architectures has to be constant and reliable, even when it is tempting to cut corners or leave out a step in evolving the control set. This means that the performance of cyber resilience control sustainment functions must be strictly overseen and managed. That oversight can be intrusive for line staff and difficult for the organization to enforce, unless the day-to-day operation of the control sustainment process meshes efficiently with the business and operating philosophy of the organization as a whole.

Because the threat environment constantly changes, confidence in the security of the control architecture must be continuously renewed. In addition, because the controls themselves are interconnected within that architecture, an appropriate level of confidence must be maintained for the entire control architecture. Typically, sustainment monitors the organization and its concomitant control architecture's ability to effectively sustain confidence in the integrity of the control set.

Sustainment monitors the control architecture's capability to successfully deliver protective service, identify and record problems as they occur, analyze those problems, and take the appropriate action to ensure the continuing security of the protected items. Monitoring makes use of defined policies, procedures, tools, and standards to continuously assess the performance of the control architecture.

If a potential threat is identified, operational sustainment prepares a formal response or creates an appropriate new control set. Sustainment monitors and assures any changes to a control or control array and facilitates proper reintegration of the altered control behavior. Management must make certain that the sustainment function is properly resourced, since it is that function that ensures continuing security. Monitoring and assurance involve four types of assessments:

1. *Corrective monitoring and assurance*—involves identifying and removing potential vulnerabilities and correcting actual errors.
2. *Preventive monitoring and assurance*—involves the identification and detection of any latent vulnerabilities.
3. *Perfective monitoring and assurance*—involves the improvement of performance, dependability, and maintainability.
4. *Adaptive monitoring and assurance*—seeks to adapt the control array to new or changed environmental factors.

It is necessary to monitor the ongoing functioning of the control architecture within the real-world operational environment of the organization. That is because threats can arise at any point and can represent a range of unanticipated dangers. Therefore, this monitoring must be done on a disciplined and regularly scheduled basis.

Identified threats, vulnerabilities, and violations are recorded and reported using a formally defined problem reporting and problem resolution process. The operational monitoring process is initiated by the

1. documentation of an overall concept of control operations;
2. preparation of an operational testing plan;
3. preparation of a Disaster Recovery Plan.

Exceptions to the intended operation of the controls must be identified and reported through a standard and disciplined process. The process must be both standard in its procedure and fully documented. The process also must be well understood within the organization. However, it is also necessary to ensure that operational monitoring is carried out in the most effective and efficient manner. Consequently, the operational assurance process itself must be continuously monitored.

Translating Monitoring Information into Action

If the monitoring process is effective, it will identify anomalies or other unforeseen events that must be addressed in order to maintain proper security control over the protected space. In general, that amounts to the identification of impacts and impact options for the prospective change. Impacts on the control architecture as well as organizational goals must be understood and characterized. Implications of any prospective change must be fully examined and documented and communicated to the appropriate decision maker for authorization.

In order to facilitate this, the results of the analysis process are formally recorded and maintained in a permanent repository. To implement changes properly, it is essential to understand the evolutionary process of all the components and the consequences. That degree of understanding requires knowledge of all aspects of the control architecture and the affected controls. Thus, it is good practice to

1. describe the precise nature and implications of any prospective change;
2. develop an overall response strategy;
3. identify control elements to be modified in the existing control array;
4. develop a specific change strategy for each element using good practice;
5. identify and ensure interface elements affected by the modification;
6. assure recommended response strategy against relevant security and safety policy.

Attaining this level of understanding requires a comprehensive and detailed impact analysis. This analysis should be based on a formal methodology, ensuring a complete and unambiguous understanding of all operational implications for the proposed control change, its requirements, and its associated architecture. Therefore, for each control remediation option to be considered, it is recommended that analysts should

1. identify the violation, exposure, or vulnerability;
2. perform a comprehensive risk identification;
3. identify the scope of the violation, exposure, or vulnerability;
4. provide a formal statement of the criticality of the violation, exposure, or vulnerability;
5. document all feasible options for remediation;
6. assess the impact of change on security and control architecture;
7. estimate implications as they impact the policy and procedure infrastructure;
8. estimate the impact of change on the Disaster Recovery strategy;
9. estimate resource requirements, staff capability, and feasibility of the change.

To support explicit decision-making on how the organization will respond, the body of evidence must be communicated to the designated approving authority for authorization of the change. The aim is to maintain the security and integrity of the control architecture throughout its useful lifetime. The reality is that over that lifetime a significant number of new vulnerabilities, which might threaten that security will be discovered. Thus, there has to be a formal mechanism responsible for executing and overseeing rational change management.

Chapter Summary

The design stage creates the requisite controls for ensuring the organization's priority assets. Therefore, the design phase satisfies the explicit control objectives for each critical asset. Design prioritizes the business' security goals and then implements a full set of targeted control behaviors to achieve the highest priority objectives. Design analyzes and assesses the resultant control set to ensure that the subsequent infrastructure as a whole satisfies the critical purpose of cyber resilience. The customary outcome of the design process is a formal and fully documented cyber-resilient architectural solution.

Architecture is an industry term that describes a comprehensive set of fully interacting and conceptually discrete array of technical and managerial controls. A control architecture embodies specifically tailored technical activities and tailored business procedures. These comprise a coherent system. Every architecture must be implemented to fit contextual requirements. This necessitates a comprehensive process to understand the precise form of the threatscape as well as the specific life cycle requirements of the active system that the architecture represents.

The first step in that process is to define precisely what the control architecture will encompass and protect. As with any other type of security approach, that process begins with establishing precisely what assets will fall within the general boundaries of the security scheme and what will be intentionally left out. The designation of a protection perimeter involves defining what the system will defend. No matter how elegant the solution, the protection scheme is useless if it is not feasible. Consequently, a central part of the process of planning and deploying a practical cyber-resilient perimeter is the assessment and trade-off between known threats versus the financial and staff resources required to secure a given item.

A decision maker uses the asset prioritization process to maximize security resource use and to optimize the overall resilience of the protected space. Even with precise knowledge of the form of the prioritized asset base, the borders of the protection perimeter are subject to ongoing refinement as threats and resources evolve.

A control specifies the precise collection of behaviors required to counter a credible threat to a given object. In practical terms, those objects are usually vital business assets or critical information. The development of a substantive architecture of controls is a formal, strategic organizational planning activity. Therefore, it must be initiated and overseen in the same manner as any other business venture.

Before any protection scheme can be developed, the form of the asset must be known and categorized. This is not a trivial exercise. It is prerequisite for subsequent control formulation because it establishes the "day one" state of the organization's entire array of potential items for protection. In most instances, control behaviors are based around and embody best practice. That is, the control is designed to operate in a way that is generally considered to be the best practical way of achieving a security purpose.

Asset valuation is important to control design and deployment because there is a direct correlation between the resources that will be deployed to establish a given level of control effectiveness. Operational factors that effect that consideration include answers to the following two questions: (1) What is the level of value of each given asset to be controlled? (2) What is the exact resource commitment that will be required to assure it? Thus, the economic and operational values of each asset must be known in order to decide what priority to assign it. Since the design of the controls will be based on the degree of rigor justified by the asset's inherent value, the actual process of designing and implementing the control can only be determined once its priority is known.

Each of the components in the architecture must be assured by substantive control measures. That implies that the risk mitigation actions must be turned into a planned and intentional set of standard operating procedures. Those procedures ensure that each identified risk is addressed by a robust set of control behaviors. Therefore, the first step in that process is the visualization of specific risks and their concomitant attack vectors, for each of the priority components. That is then followed by the deployment of a substantive control that is aimed at addressing each of those parts.

Detailed design descriptions provide a precise description of what each control is supposed to do as well as which of the other controls it will interact with. In the architectural design phase, the organization also produces a detailed set of testing and review qualification requirements for each priority component in the design. And, once these are known the schedule for testing each component is laid out. Ultimately, all of the deliverables in this phase can be evaluated based on the representative criteria of traceability, external and internal consistencies, appropriateness of the methodology, and feasibility.

The individual controls that will populate the cyber-resilient control architecture are deliberately designed entities. Whether the control is a policy, a specific recommended action, or an electronic operational feature, each of them is designed to specifically address an identified risk or threat. Thus, control design occurs once the threat picture is established. The controls themselves are behavioral in that they specify a specific action to be taken or programmed response that will occur, given the fulfillment of certain criteria. These actions or responses are preprogrammed and must happen as intended by the design.

This requires an item-by-item evaluation of each information element and its attendant threats, to define and formalize the requisite actions to be taken.

That assessment evaluates every possible threat vector and then deploys a set of practical actions to ensure that no undesirable contingencies occur. This design and deployment planning is done for every information element in the asset base.

Because the overall goal of this step is to ensure the correctness of the final product, each control must be tested against a set of assigned criteria that have been designed to assess, without any ambiguity, control performance. This level of specification ensures a systematic evaluation of each critical component in the cyber resilience scheme. As such, each control must have a behaviorally observable result associated with it.

Finally, it should be clear that the essential condition for the successful creation and sustainment of an effective cyber resilience control architecture is for all members of the organization to have common understanding and buy in for the behaviors that are expected of them. There is no way that the organization can hold an individual accountable if he or she was not given adequate information on how to behave. Therefore, to some extent, the success of the system centers on establishing and maintaining effective communication. All participants must understand the rules of behavior, because cyber resilience control schemes are complex and subject to interpretation by the participants. Participant understanding must be controlled to some extent to ensure consistent outcomes.

The purpose of the control design process is to develop and maintain an explicit model of the control architecture that includes precise specification of the control architecture's required behavior. This specification is then used to operate and sustain the practical control system. In many respects, control design is the single point where the organization can intentionally ensure a high level of security and integrity.

The organization installs and tests each control behavior in accordance with a specified implementation plan and ensures that each installation satisfies the overall requirements of the cyber resilience strategy. In that respect, the implementer must develop and document a set of tests, test cases, including inputs, outputs, test criteria, and test procedures to assure the practical effectiveness of each installed control.

This is all done by plan. That plan should include, as applicable, a description of the control context/environment, a rationale for the planned behavior based on that description, plus all information and analysis procedures to ensure the effectiveness of control performance. The plan should identify the larger threat protection requirements that have been addressed by each control and ensure that all requisite control requirements are assured by some form of qualification testing. The plan should also determine the appropriate timing and sequence to begin installation and provide details regarding all testing and assurance procedures, staff roles and responsibilities, and the documentation that must be prepared and delivered to ensure hand over to the operational use of the system.

The maintenance of a secure control architecture involves the development, deployment, and continuous maintenance of the most appropriate and security response to threats and changes. To accomplish this, standard security and control processes must be planned, designed, administered, and maintained. The aim is to ensure that effective leadership vision, expertise, and the correct technology solutions are available to assure effective security and control of the cyber resilience control architecture.

Standard procedures must be in place to continuously assess, integrate, and optimize that security architecture, and the security architecture must be maintained consistent over time. Therefore, future control assurance trends also have to be evaluated to define the long-term architectural strategy. Emerging threats and security trends also have to be assessed as they might impact the evolution of the security architecture, and resources have to be developed and made available to ensure the long-term effectiveness of the control scheme.

The long-term assurance of the control architecture is specifically defined and formally implemented through a Sustainment plan. Commitment to this plan must be rigorously maintained throughout the life cycle. At the minimum, this plan should specify the way changes to the control baselines will be managed.

The purpose of the sustainment function is to establish and maintain the integrity of the architecture of controls that has been established for that particular organizational asset. Control sustainment does this by developing a coherent strategic oversight and change management process. The specific aim of this process is to ensure the completeness, consistency, and correctness of each of the control elements in the formal control architecture. Consequently, once the control architecture has been established, the sustainment function controls the modification and integration of every baseline item. Sustainment also records and reports on the status of each of these items. And, it maintains a record of the monitoring and oversight functions.

Because the threat environment constantly changes, confidence in the security of the control architecture must be continuously renewed. In addition, because the controls themselves are interconnected within that architecture, an appropriate level of confidence must be maintained for the entire control architecture. Typically, sustainment monitors the organization and its concomitant control architecture's ability to effectively sustain confidence in the integrity of the control set.

Sustainment monitors the control architecture's capability to successfully deliver protective service, identify and record problems as they occur, analyze those problems, and take the appropriate action to ensure the continuing security of the protected items. Monitoring makes use of defined policies, procedures, tools, and standards to continuously assess the performance of the control architecture.

If a potential threat is identified, operational sustainment prepares a formal response or creates an appropriate new control set. Sustainment monitors and assures any changes to a control or control array and facilitates proper reintegration of the altered control behavior. Management must make certain that the sustainment function is properly resourced, since it is that function that ensures continuing security.

Keywords

Architecture: The design and implementation of an underlying framework of processes

Best practice: A set of lessons learned validated for successful execution of a given task

Baseline: The collection of a set of objects all commonly related by application or purpose

Control design: Specification of behaviors of a protection measure or measures

Control evaluation: Formal testing or reviews of a control or control set to confirm correctness

Control performance: The operational results of control operation within a given environment

Controls: A discrete set of human or electronic behaviors, set to produce a given outcome

Control architecture: The logical array of well-defined actions deployed to enact a solution

Critical asset: A function or object that is so central to an operation that it cannot be lost

Cyber resilience: Assurance of the survival and continued operation of critical assets

Operational sustainment: Planned actions meant to maintain a control architectural purpose

Operational testing: A planned set of tests and reviews designed to assure performance

Reliability: Proven capability to perform a designated purpose over time

Resilience management: Formal oversight and control of resilience actions of an organization

Strategic planning: The process of developing long-term plans of action aimed at furthering and enhancing organizational goals

Chapter 6

Control Assessment and Assurance

At the end of this chapter, the reader will understand

1. the importance of testing and assurance in the maintenance of a secure organization;
2. the various phases of the testing and assurance process;
3. strategies for implementing a continuous assurance operation;
4. the issues associated with control assurance in an operational environment;
5. standard tests and reviews for control assurance;
6. how review and audit contribute to cyber resilience.

The Need for Reliable Assurance

Because they ensure the organization's priority assets, the control set in a cyber-resilient system must be thoroughly and consistently inspected and tested. The controls that make up that system are real-world entities, which are subject to failure and irrelevance. Therefore, the general process for ensuring their reliability is a continuous assurance function, which is capable of characterizing the explicit level of performance of the current control architecture against the organization's assurance goals. The aim of that process is to be able to say with reasonable certainty that the aggregate control set for any given cyber-resilient protection scheme is effective, given the strategic aims of the organization.

In general, this is an assessment function that is comparable to the systematic monitoring and adjustment processes that most organizations employ to assure the effective performance of their various key functions. Operationally, the evaluation and assurance process should operate within a defined reporting and decision-making organizational structure. Because the overall purpose of any assurance process is to produce a trustworthy state of organizational operation, the outcome of the Assurance Phase is continuous guarantee of control correctness.

With respect to the continuous assurance of control effectiveness, it is important to note that systematic security control measurement and evaluation activities are routine responsibilities of overall cyber resilience risk management. The security control assessment process provides vital information that is used by management to make long-term strategic decisions regarding security directions. But assessments also support many other security, risk, and control resource management activities.

Nonetheless, assessments also play a key role in operational management of the security function and other day-to-day security management activities. Control assessment processes are executed at a much more pragmatic and application intensive level than the processes that are associated with control formulation and development. Depending upon individual organization security goals, security assessments may be performed on a planned and scheduled basis within the normal operation of the cyber resilience process. During these assessments control, developers and implementers work collaboratively with line managers and stakeholders on specific efforts to support the ongoing operation of the business. That includes such fundamental operational activities as new product development planning and general security protection activities such as vulnerability scanning, security control validation, and routine new control integration, and regression testing.

Since one of the primary objectives of security control assessment is the identification of potential weaknesses or deficiencies in the control array, as implemented, organizations must conduct security control assessments during the threat assessment and strategic planning phases of routine operations. The aim of these activities is to confirm the proper and correct performance of the control set and its configuration as it addresses known threats in the current threatscape.

For any organization with a formal compliance requirement, periodic control assessments of their operational systems also help to satisfy the requirements of any specific legal mandates. Finally, continuous assessments help to better develop and implement continuous monitoring requirements for the long-term control development process.

As we said in the last chapter, organizations will generally utilize some form of standard, generally accepted model for control architecture as a means of facilitating the explicit creation and subsequent assessment of controls. This will be discussed in much greater depth in Chapter 8. But, in general, these requirements will always apply to federal agencies who must implement NIST 800-53. Private sector industries may use everything from ISO 27000 to information systems audit and control association (ISACA)'s control objectives for information and related technologies (COBIT) model to structure practical responses to legal requirements like Sarbanes–Oxley or Health Insurance Portability and Accountability Act (HIPAA). NIST SP 800-53A is particularly helpful in meeting retirements because it provides a comprehensive set of recommended control assessment procedures, which it presents in a well-defined and commonly accepted format.

Evaluating Control Behavior

The overall aim of any standard assessment process is to review, inspect, audit, or test every individual behavior in the control array. For instance, every NIST 800-53A based control assessment process is based around one or more assurance objectives that specifically state what the evaluators must determine in order to certify the effectiveness of each control in the organizational cyber resilience architecture. Every one of these assessment objectives is associated with assessment methods and assessment objects that define how the evaluation team will subsequently assess the control and what each assessment activity should focus on.

As in most organizational projects, the process is initiated by a formal assessment planning process. The purpose of that process is to determine the scope of the overall assessment process and choose the appropriate methods as well as target the explicit assessment objects for each control. The specification of how each general assessment effort is applied can vary from examinations of individual control components and behaviors to comprehensive evaluations of the overall performance of the control array. As we said, this is generally dictated by the requirements specified in the generic control model that has been selected to achieve a level of assurance consistent with the minimum security requirements of the organization.

Whatever model is selected, the formal guidance that the organization employs in order to fully plan and perform the assessments themselves ought to fully specify the behavioral requirements at the appropriate level specified in order to ensure that control performance has been adequately examined, sufficient evidence gathered, and tests performed. The actual management boots-on-the-ground during the assessments will utilize these specifications to plan the resource commitments as well as the degree and nature of the evidence required to draw a proper conclusion about each individual control. This decision will also dictate the level of detail required to adequately document control performance for managerial actions.

Organizations have considerable flexibility in adapting security control assessment procedures to suit their particular threat environments. This is done during the operational planning phase of any given assessment. Just as the process of security control design and development allows the organization to tailor their security control architecture to reflect the demands of the existing threat environment, the capabilities of their security personnel and available resources have to be tailored to the situation.

The aim of the planning and evaluation phase is to establish predefined objectives along with a set of tailored assessment methods and assurance cases that have been specifically targeted on the assurance requirements of the organization for that particular assessment project. The security control assessment process tailors the formal assessment procedures to match the precise requirements of any given assessment activity.

In those instances, the focus and level of rigor of the assessment is driven by a generic commonly recognized strategy and well-defined test plans for the security control array as a whole. That allows the assessment of each individual control to be conducted in a uniform and consistent manner. The overall aim is to eventually be able to provide a set of reliable and repeatable results.

The actual evaluation is driven by the assurance case. The beauty of a suitably articulated assessment case is that it ensures the alignment of the assessment effort with the larger, strategic level organizational goals. That can reduce the time and level of resources required to execute a proper security control assessment process.

These cases are defined in advance as part of the operational planning activity. They itemize and explain the rationale and attendant set of specific steps that the assessment team will take in order to gather the evidence that is necessary to evaluate the controls and any requisite control enhancements. The assurance case is then fitted to a relevant assessment method. Each assessment case is developed from a specific planning perspective. That perspective encompasses the appropriate and commonly accepted business purposes that would motivate and guide any standard organizational assessment. However, the case may be adjusted for any other identified organizational or threat-specific requirement.

The security control assessment plan defines the exact means that the organization will utilize to perform the specific activities of the assessment. In general, every process will attempt to verify the effectiveness of the target security controls. The precise actions that will be taken to accomplish that verification are documented in the operational steps of the plan. Those activities can include any form of standard testing, review, or audit of evidence produced by real-world control operation. Evidence gathering includes anecdotal interviews of staff who are knowledgeable about system operation and unit and integration testing of the relevant controls to determine whether they are operating as specified in the assurance case.

Operational Assurance of the Control Architecture

The control architecture assurance process involves any form of assurance activity planned to establish and enforce a cyber-resilient organizational state. Generally, this means that an explicitly planned set of testing and review activities are deployed, focused on determining how correctly and effectively each control adheres to the requirement and design specifications used to design it and the criteria for performance laid out in the design.

As we said, performed properly, the assurance of the control architecture is done in parallel with the practical control development process, not after the architecture has been created. Thus, the assurance is iterative as the architecture evolves. The aim of this assurance activity is to enhance overall management insight into the risk status of the organization as well as ensure that the controls and the control architecture comply with the explicit performance, schedule, budget, and control integrity requirements of the organization's specific cyber resilience architecture.

Control architecture integrity levels comprise a range of critical control properties that must be assured in order to maintain risks within acceptable limits. Shown in Figure 6.1, the criticality properties include

1. the frequency of occurrence and impact levels of threats in the threatscape;
2. implicit and explicit safety requirements;
3. implicit and explicit security requirements;
4. the degree of control complexity;
5. control performance factors;
6. reliability factors.

The requirements are normally defined early in the architecture development process. These requirements can range depending on the organizational context. Obviously, mission critical controls require a high degree of integrity, which implies

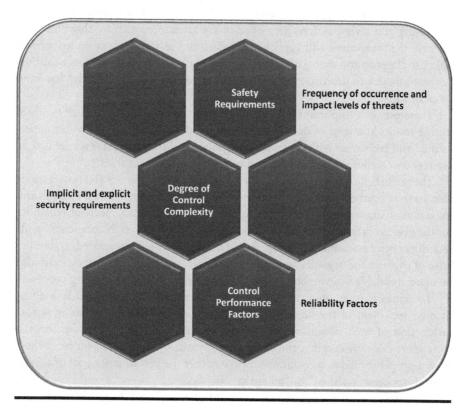

Figure 6.1 Criticality properties.

a large and rigorous number of tasks to address these factors. Whereas, a simple, noncritical business organization, like a party store, will require much fewer and less resource intensive controls. But in every case, some form of control architecture is required.

Assurance of that architecture is documented in the Control Architectural Assurance Plan. That plan coordinates all the diverse activities of the control assurance process. The aim is to ensure that every validation and verification process is properly related to all the relevant aspects of the project. That plan specifies the requisite assurance process and attendant resources necessary to perform all the activities and tasks needed to provide trusted cyber-resilient processes.

Establishing a Regular Organizational Testing Process

Testing is an objective, often hypothesis-based assessment of the status of a given control, control array, or control architecture. Because its outcomes are scientifically derived, testing is a key part of assurance. The testing process should always follow a defined test protocol, which is spelled out in detail for every control to be tested. The assessment itself tests or examines the behavior of each control by performing the testing and review actions prescribed in the assessment plan for that object. In essence, the assessment will explicitly examine and test every relevant assessment object in the plan and then gather and document all the necessary evidence to allow decision makers to determine whether every stated assessment objective has been properly satisfied.

Therefore, it is important that the assessment team document security control testing results at a level of detail appropriate for the type of assessment being performed and consistent with organizational policy and the expectations specified by managers or senior executives who will review the assessment results.

The tests themselves must be supported by objective outcomes and observations that serve as hard evidence of the target performance of each assessed control, and must demonstrate completeness, correctness, and an acceptable level of reliability for the evidence that is presented to support a given conclusion. Nonetheless, while security control assessment findings should be objective, evidence based, and indicative of the way the organization implements each security control, they must also be understandable by everybody who is involved in the decision-making process.

To reinforce that objectivity, every assessment plan should include a documented statement of the outcomes that would be considered adequate to ensure satisfaction of each stated assurance goal. Because the aim of the assessment is to support decision-making, the evidence presented by the assessment team must draw a properly validated conclusion as to whether the performance goal of a given control has been "satisfied" or "other than satisfied."

The assessment team will render a conclusion of satisfied if there is substantive evidence that the control meets the assessment objective. A finding of other than satisfied indicates that the evidence found is insufficient to meet the assessment objective. To justify each other-than-satisfied conclusion, the assessment team must document the aspects of the security control behavior that were deemed unsatisfactory or were unable to be assessed. That must be explicitly stated for every assessment goal.

Then the evaluators must fully describe how the control, as assessed, differs from expectations. It is important to note that while the discovery of weaknesses or deficiencies in a control's implementation may result in an "other-than-satisfied" conclusion, there might be mitigating circumstances that would make the control acceptable, for instance, situations where the assessment team was not able to obtain sufficient information to draw an objective conclusion about the performance of the control.

Upon finalization of the testing report for a given evaluation project, the organization has to begin a formal remediation process for each of the controls that have been found to be functioning insufficiently. This requirement is satisfied through the specification of a deliberate set of well-defined remedies that the report will itemize. An organization then satisfies the requirement for due diligence by prioritizing and planning the specific actions. This final step is conducted under the dictates of a formal plan, just like any other operational project.

It must be kept in mind that the routine control tests must be customized to meet each organization's unique needs. Thus, control test planning is a critical function in control assurance. Overall control testing practices must be planned at the outset. There must be defined points where the organization pauses to thoroughly test its individual cyber resilience control requirements for their effectiveness in satisfying the organization's strategic security goals. Once the schedule for routine testing has been planned, individual control performance is evaluated to determine whether that particular item conforms to business requirements and the security policies and plans of the organization (Figure 6.2).

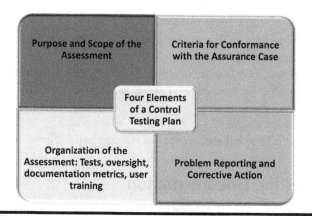

Figure 6.2   Four elements of a control testing plan.

Understandably, the control testing plan brings together the results of the prior stages of cyber resilience system development, classification and risk, control design and implementation. The control testing plan is designed to continuously ensure a uniform minimum acceptable level of cyber resilience in the organization as a whole. Thus, each plan embodies four general considerations:

1. Purpose and scope of the assessment
2. Criteria for conformance with the assurance case
3. Organization of the assessment
    a. Tests
    b. Control oversight and remediation management
    c. Evidence documentation and records collection
    d. Standards and metrics
    e. User training
4. Problem reporting and corrective action

The Control Testing Process

The first step requires the organization to delineate the specific purpose and scope of the testing process. Specifically, this delineation

1. must be particular to that organization;
2. itemize the control items to be assessed by the plan;
3. itemize the intended use for the control;
4. specify the minimum performance criteria for the controls covered by the plan.

The second consideration specifies how the testing process will be organized and conducted. It requires the organization to spell out how it will perform the control assurance performance tests including

1. control infrastructure documentation including description of all elements;
2. architectural interdependencies;
3. the specific portion of the architecture that will be addressed by the assessment;
4. the relationship between all planned assessment tasks and the schedule for evaluation;
5. responsibilities assigned;
6. the organizational stakeholder specifically responsible for each task.

Control Testing Documentation

This activity has two parts. The first part of this identifies the assurance goals that will guide the control testing process. That identification generally specifies how each control will be tested for conformance with performance criteria. The second section specifies in unambiguous terms the minimum requirements for conformance. This looks at each control from the standpoint of its purpose and minimum acceptable level of performance utilizing the

1. control behavioral specification;
2. control design description;
3. control validation and verification plan;
4. reporting process;
5. user documentation that is required for each control;
6. control configuration management process;
7. any appropriate standards, practices, conventions, and metrics to be applied;
8. how compliance will be monitored in the long term.

There has to be a commonly accepted and understood stipulation of the tests to be conducted. This document also specifies how each of these will be performed. This includes a specification of how results will be reported and non-conformances handled. It also stipulates the minimum number and types of tests and reviews to conduct. That might include

1. control requirements review;
2. preliminary design review;
3. configuration management plan;
4. control verification and validation (V&V) plan;
5. problem reporting plan;
6. specific tests to be performed and their methodologies;
7. problem reporting, which itemizes problem reporting, tracking, and resolution;
8. the specific individual in the organization who will be accountable for each role.

Test Design

Once the control testing plan has been developed and integrated with the other relevant plans, especially those for inspections and audits, it is time for the organization to define its standard testing procedures. That step involves formally establishing the control assurance testing activities themselves. This normally involves the definition and scheduling of tests. It also involves the institution and promotion of the necessary organizational discipline to ensure that control assurance testing is done properly and adds value to the organization.

Specifically control assurance testing personnel must be assured capable, and instilled with sufficient discipline to correctly perform the control tests specified.

That includes the creation and enforcement of standard testing protocols. It also includes the scheduling and conduct of routine testing activities. Ultimately, from a process standpoint, control testing must have the mandate to conduct an in-process evaluation of the entire organizational control system.

This is communicated through a documented and commonly understood strategic testing plan. The aim of this plan is to document an explicit series of activities to ensure control validation. Specifically, that plan should provide an explicit specification of the purpose, scope, and activities of the proposed control testing function. For operational purposes, control testing is defined as "the process of testing a control item to detect the difference between existing and required conditions as well as to confirm the proper functioning of the specific features of the control item."

Like any other standard activity, the control assurance process has to be intentionally designed and documented by plan based on the overall strategic goals of the organization. The elements of the cyber resilience process that must be ensured include the:

1. Architecture
2. Components
3. Interfaces
4. Operational assurance process itself

In this case, test design and execution are normally an iterative process. That is due to the fact that the threatscape is constantly changing, which affects architecture and in its turn the design of the testing process itself. Common activities in the testing of a control architecture include

1. tracing the control design to ensure that it corresponds with an organizational objective;
2. evaluation of the control architectural design to determine whether each control is operating correctly and effectively;
3. evaluation of the interfaces to determine whether the architectural design is correct and effective;
4. test plan generation and documentation for each phase of the testing process;
5. test design generation and documentation for each phase of the testing process.

Test Execution

Testing is essentially performed at several points in the control architecture development process. As we have seen, the design and implementation of controls is hierarchical. So, the control architecture evolves from an individual collection of discrete controls to a fully operational everyday architecture. Consequently, the testing process requires plans at every one of these stages:

1. Individual component control testing, which takes place during implementation
2. Control architectural integration testing, which takes place during integration
3. Routine control architecture resilience testing, which takes place as planned
4. Any form of additional certification and accreditation testing, if required

Individual component control testing—This testing is carried out to verify that the design is properly implemented and that the individual control element or a collection of related control elements function as specified in the operational space. The purpose of this kind of testing is to ensure that the control logic is complete and correct and that the control or controls function as intended and designed in the operational space.

Control architectural integration testing—These are a designed series of progressive tests and reviews in which the control elements are tested in the operational space as the resilience function is built. The purpose of integration testing is to ensure that the architectural design objectives are met. The aim is to confirm the correctness of the interfaces and the effectiveness of the controls as an integrated entity. This testing continues until the entire system is fully integrated.

Routine control architecture resilience testing—This is the persistent process of testing the fully integrated control architecture to verify that the cyber resilience control architecture meets its specified requirements. This testing is scheduled by plan and performed as a routine organizational function. Its purpose is to ensure that the control architecture, as a comprehensive entity, continues to satisfy its business overall purposes and its operational assurance objectives.

Any form of additional certification and accreditation testing—Many organizations have external requirements for security assurance. For instance, federal government organizations must meet the stipulations of the Federal Information Security Management Act. This is also true for health care organizations (HIPAA) and publicly traded corporations (Sarbanes–Oxley).

In all those instances, and many others, there is a formal testing requirement that is aimed at determining whether or not a system satisfies a stated set of standard criteria. The goal is to provide assurance that the organization has complied with some legal or regulatory requirement for information protection. The tests assure that the security requirements of the regulatory law or standard are met and that all the requisite controls are in place.

The specifications for performing all testing activities are communicated by a formal testing plan. This plan is a formal organizational-level document. It spells out the scope, approach, resources, and schedule of the testing activity. It stipulates risks and allocates resources. It explicitly states who will perform the tests, the accountable stakeholders, and how the organization will transition from one phase of testing to another. There is an explicit estimate of the number and types of test cases that will be required, the duration of the testing period, and the criteria that will indicate satisfactory completion.

Test Design

Once a testing plan is completed and accepted by executive management, the tests themselves have to be designed. In order to be effective, every testing activity needs to include an explicit statement of: (1) the required features to be tested, (2) the criteria for successfully passing the test, (3) the criteria for control performance, (4) the degree/extent of documented security performance required including the itemization of all known threats, and (5) the criteria for secure operational interface performance.

Test cases and related procedures are developed to fully explore these questions. Test cases specify the actual input values and expected results for each input. The goal is to exercise the control component's logic and to set up testing scenarios that will expose errors or produce unexpected results. The aim is to undertake the smallest number of cases possible that will still meet this goal.

Test procedures: Once the test cases have been defined, the next step in the process is to develop the explicit test procedures. These procedures identify all the practical steps that will be required to demonstrate successful and effective control functioning. Each procedure is designed to satisfy the objectives of the specific test cases.

Test execution: Then the test procedures are executed as stipulated. The actual execution process begins at the individual control component testing level and proceeds up through the hierarchy of testing levels to any required certification accreditation testing.

Ensuring the Reliability of the Control Architecture

A control can be trusted if it can be shown that it will behave in a predictable way and generate any result or output that could cause harm. The ability to trust control operation is normally termed "control reliability." That term is reinforced by a broad class of reliability assurance procedures that are undertaken to ensure that a control is safe. For instance, a sophisticated range of assurance methodologies might include such things as, and shown in Figure 6.3:

1. Fault tree analysis (FTA)
2. Formal methods (particularly early life cycle)
3. Criticality analysis
4. Impact analysis
5. Control safety analysis

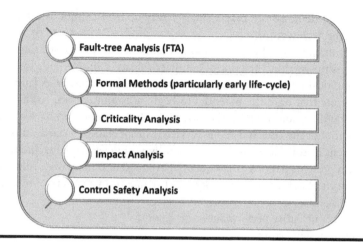

Figure 6.3 Assurance methodologies.

Conventional control assurance activities should be undertaken early in the architectural design phase of the control life cycle. That is due to two important reasons: cost effectiveness and feasibility. First, catching problems in the early stages of the design phase ensures the most cost-effective remediation. Second, it is important to know whether the controls that will populate a given architecture are feasible, prior to undertaking the actual implementation process. Feasibility is a pure business consideration.

However, the downside of performing nothing but complex control reliability analyses is that the resultant controls will not achieve the original design goals as implemented, and therefore will not achieve the desired level of cyber resilience. Therefore, the control reliability assurance process focuses on the practical behaviors of the controls and the interrelationships between the control activities. These both must be shown to be sufficient to be able to trust the control array to provide the desired protection. Consequently, the aims and purposes of the overall cyber resilience system must be completely and unambiguously understood before control reliability testing can be tackled.

Using Fault Trees to Enhance Understandability

Control reliability assurance can be viewed as a subproblem of ensuring overall cyber resilience. The goal of a fault-tolerant design is to ensure that if the domino effect begins, there are some dominos in the line that are strong enough, that even if they are hit by a falling neighbor they do not fall. In other words, at least some of the controls have to be made resilient enough to ensure that one domino toppling does not bring down the entire line of dominos. Thus, the goal of control reliability assurance is to demonstrate that a single faulty control cannot affect the overall security of the architecture.

As a result, perhaps the most common approach to demonstrating the exact interrelationships within the control architecture is via simple fault trees. A fault tree is a graphical representation of events in a tree-like structure. A fault tree captures all events that have the potential to begin a domino effect. In that analogy, if the last standing domino falls, the hazard can be considered to have occurred.

So, the root of the fault tree represents a hazard or class of hazards. It is the final harmful event in some sequence that hopefully will never happen. The remainder of the tree represents parallel and sequential events that potentially could cause the hazard to occur. Such potentially risky events come from a variety of sources:

1. Incorrect or faulty performance of the control
2. Logical design errors
3. Human error
4. Malicious attacks
5. Organizational breakdowns or failures

Fault trees are conceptually simple models. They are composed of events and logical event connectors. Each event node's children are the necessary preconditions that could cause that event to occur. These conditions can be combined in any number of ways using if-then-else logic of failure associated with each node.

The use of fault trees in the design evaluation can also help determine the non-critical control components versus the critical ones. This is important because the integrity of the critical components in the control architecture must be maintained in the face of any form of attack or exploitation attempt, not just in normal operation.

Reliability Metrics

One notional approach to ensuring the integrity of critical components is to objectively analyze all the possible interactions between critical components versus the noncritical ones. Thus, the performance characteristics of controls are captured and described by means of quantitative reliability metrics. The growing acceptance of a metrics-based approach has driven the growing use of explicit reliability metrics in any form of control assurance.

Reliability metrics embrace all types of quantitative algorithm or function that can be used to obtain objective evidence of the performance of an entity, such as a control. The actual numerical value produced by a metric is called a measure. There are two general classes of metrics:

1. *Management metrics*, which assist in the management of the control architecture development and sustainment process

2. *Performance metrics*, which characterize and predict the discrete behavior of the control

In general application, management metrics are used to characterize the performance of any form of production or manufacturing activity. Management metrics are collected and analyzed throughout the cyber resilience control development and sustainment process. These metrics are easily understandable by decision makers because they can be plotted using bar graphs and histograms or as part of statistical process control. The plots can be used by management to identify the control activities that are most successful and the ones that are the most error prone. Using this information, managers can develop mechanisms to prevent the recurrence of similar errors, suggest procedures for earlier detection of faults, or make improvements to the control development process.

Performance metrics are not the same as management metrics. Performance metrics are used to characterize specific characteristics or behaviors of a control, such as number of attacks prevented or successful exploitations. They are used to characterize the day-to-day behaviors of the control set. Normally, these measures lack a quantifiable basis for comparison. Therefore, they are often interpreted and used by comparison with prior plans, similar projects, or similar components within the current control array.

Making Controls Reliable

Controls are reliable if it is impossible, or at least highly unlikely that control behavior could ever malfunction in a way that would cause an undesirable event. Examples of undesirable events include loss of information, loss of physical property, or catastrophic loss, such as total system failure.

Control reliability assurance usually refers to a broad class of design, implementation, and operational activities, methods, or tools that can be utilized to ensure that the organization will be safe from an undesirable event. This includes such techniques and methods such as:

1. Criticality analysis
2. Impact analysis
3. Control reliability analysis

The downside of applying control reliability analysis only to the design is the possibility that the implementation or the ongoing operation of the control will not reflect the original design intent, and therefore will not provide adequate protection. When control reliability is applied to actual operational controls, it is possible to demonstrate that the control functions as intended on an ongoing basis.

As we said earlier, perhaps the most common approach to demonstrating control reliability is via fault trees. But there are other commonly accepted methods. One common method employed for ensuring operational security in the control array is to build the controls in a modular fashion. In simple terms, this means that critical control functions are isolated from modules that do the actual work of the organization. For instance, controls that assure correctness of a bank account are isolated from the process of depositing or withdrawing funds.

Modularity also plays an important role in separating critical assurance functionality from noncritical functionality. For instance, the access control system is isolated from other systems that perform account assurance such as balance checking. In simple terms, the aim is to ensure that the behavior of noncritical control components do not have undesirable or unpredictable effects on the critical control components. The integrity of that modular relationship must be maintained. Therefore, one approach to ensuring integrity is to analyze all possible interactions between critical control components and the rest of the control architecture. Unfortunately, given the complexity of most cyber resilience solutions this is difficult. Therefore, some form of measurement-based assurance is required.

Measurement-based assurance simply means that confidence in a given control can be assured through objective measurement. This is based on the availability of a range of standardized metrics that allow the organization to assess the performance of a given control array or architecture at a given point in time in a given situation. Over the past 20 years, there has been a growing trend toward quantitative metrics.

A metric is the term for a mathematical definition, algorithm, or function used to obtain a quantitative assessment of a product or process. The actual numerical value produced by a metric is called a measure. Thus, for example, defect identification and removal is a metric, but the actual removal rate for a given organizational assessment process is the measure. There are two general classes of metrics:

1. Management metrics, which assist in the management of the control development and sustainment process
2. Security metrics, which are predictors or indicators of the general effectiveness of a given control, control array, or control architecture

Management metrics can be used for controlling any form of process or activity. They are normally used to characterize resources, cost, and success rates. Security metrics are used to estimate the general performance and/or effectiveness of a control, control array, or control architecture. It should be noted that there are a number of metrics that can be used to evaluate both management and security concerns.

The primary disadvantage of a metric's approach to security is the lack of a standard assessment scale to ensure consistent interpretation of the data. For instance, it is hard to decide what a given number of security weaknesses means. This is

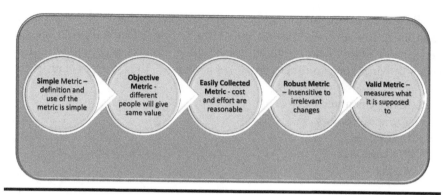

Figure 6.4 Five metric characteristics.

particularly true of metrics for subjective qualities such as "reliability." Also, while metrics are mathematically based, most have not been proven correct.

Since there are any number of possible metrics to evaluate control performance, managers must designate the specific criteria they will apply in their own cases. Ideally, a metric should possess five characteristics as shown in Figure 6.4:

1. Simple—Definition and use of the metric is simple.
2. Objective—Different people will give same value.
3. Easily collected—The cost and effort are reasonable.
4. Robust—Metric is insensitive to irrelevant changes.
5. Valid—Metric measures what it is supposed to.

Metrics are the product of formal organizational processes. These processes are characterized by three types of activities: tests, reviews, and audits. Two of those, the reviews and audits, are people powered in that they involve expert opinion, which might be subjective, but which allows the organization to both evaluate the product as well as educate the people in the organization. These are perhaps the most important mechanisms for evaluating the assurance process, since they are by far the most frequently done in the real-world assessment of security performance.

The Control Architectural Review and Audit Processes

Reviews and audits differ from tests in that they are based on expert opinion, rather than objective evaluations of performance with respect to a defined set of testing goals. However, a review or an inspection is much easier to perform, and they can take place at any stage in the control architecture development process. They can also target both controls and the process that they operate in.

There are five types of inspections, reviews, or audits that might be performed to ensure control architectural correctness. Each of these review types involves the performance of well-defined procedures to achieve the assurance objectives. These review activities apply to either a control or its context. The five types of review are:

1. Management review
2. Technical reviews
3. Inspections
4. Walk-throughs
5. Audits

Management Reviews

Management reviews ensure the correctness of management plans, schedules, requirements, control, and architectures. They support decisions about control performance, architectural effectiveness, emerging threats, corrective actions, allocation of resources, and sustainment of a requisite level of security and integrity. They directly support the management personnel who have assigned responsibility for the system. They are meant to discover and report variations from plans and/or defined procedures. They primarily focus on planning and/or corrective action artifacts.

Management reviews are carried out in support of the organization's management personnel. A management review normally involves the following potential management roles:

1. *Decision Maker*—the manager(s) who are being supported by the review
2. *Review Leader*—the single designated individual who is responsible for the conduct of process and accountable for the results
3. *Recorder*—the single person responsible for documenting a permanent record of the outcomes of the review process
4. *Management Staff*—the management personnel in the affected organizational unit accountable for the collection, preparation, and analysis of results
5. *Technical Staff*—the technical personnel in the affected organizational unit accountable for the collection, preparation, and analysis of results
6. *External Representatives*—where appropriate

Management reviews can consider the following evidence:

1. Statements of objectives
2. The individual control itself
3. The architectural plan
4. Control array status relative to plans

5. Any identified anomalies
6. Control and contextual organization procedures
7. Resources, status reports, and pertinent regulations

Management reviews should be scheduled as part of initial control architectural planning. The actual timing is usually tied to milestones and terminal phases. This does not exclude any ad hoc reviews that might be scheduled and held for the purpose of reacting to control breakdown or malfunction, or for functional management reasons. They can also be held at the instigation of an outside agency, for instance a governmental entity where compliance must be certified or even a potential business partner or customer.

Formal reviews are the products of an intentional strategic planning process. The aim of that process is to ensure that adequate funding and resourcing is available to carry out the assessment tasks, that the task assignments are realistic, and that the overall control development schedule and inspection points are feasible, and finally, that the reporting and feedback lines are well-defined and commonly understood.

The planning formulates an evaluation team and assures that team is composed of competent personnel. Each member of the team is given an explicit assignment of their roles and responsibilities and specific tasks within the review process. Following that, all pertinent materials are distributed to the review team, and training and orientation occurs where that is necessary.

Once the review materials have been distributed and the team is properly oriented, the individual reviewers begin a preliminary analysis of their parts of the assignment. That boundary is normally delimited by and focused on individual control performance. The reviews are objective. Each individual member of the review team documents and classifies any deviations from expectations regarding performance against the objectives for that control. This can also lead to conclusions about overall control system performance with respect to the strategic goals of the organization.

Any identified anomalies or nonconcurrences are reviewed with the organization's decision makers, and decisions are made about remediation where appropriate. Those decisions are then captured in formal documentation of action items and recommendations.

Once the review is complete, there is a need for formal follow-up and closure. In essence, the organization has to certify that any action items arising out of the review process have to be closed out either by generation of documentation that certifies the remediation or by a decision to accept the risk. The documentation should include

1. a unique review identifier and reviewer names;
2. review objectives for that review;
3. control(s) reviewed;

4. evidence input to the reviews;
5. itemization of action item status;
6. defect, vulnerability, and anomaly list.

Technical Reviews

Technical reviews produce empirical evidence of control performance. They support decisions about whether the control architecture and individual controls conform to specifications of performance that are itemized in any relevant regulations or plans. The aim is to objectively determine that the control architecture has been correctly implemented or changed.

Technical reviews are carried out for the purpose of supporting the organization's strategic technical direction through the provision of information to the technical and management personnel who have direct responsibility for control performance. The primary focus of a technical review is on the control and what it produces in terms of actions or outputs. These reviews are meant to discover and report all defects and anomalies in the control operation, whether it is being initially implemented or if it is an existing control that is being changed.

Technical reviews are normally carried out by the organization's technical staff and its management. The following potential roles are involved in a technical review:

1. *Decision Maker*—the manager(s) who are being supported by the review
2. *Review Leader*—the single designated individual who is responsible for the conduct of process and accountable for the results
3. *Recorder*—the single person responsible for documenting a permanent record of the outcomes of the review process
4. *Management Staff*—the management personnel in the affected organizational unit accountable for the collection, preparation, and analysis of results
5. *Technical Staff*—the technical personnel in the affected organizational unit accountable for the collection, preparation, and analysis of results
6. *External Representative Technical Staff*—where appropriate

Technical reviews are normally objective driven. That is, there is a precise specification of what is to be examined and the expected outcomes directing the technical review process. Consequently, the preparation of a specific statement of objectives is the first step in the technical review process that includes a listing of

1. the precise control elements being examined;
2. inspection/testing schedule and testing protocols to be applied to each element;
3. target anomalies and defects and reporting protocols;
4. relevant regulations also have to be considered.

Technical reviews can be both planned and ad hoc. However, the primary purpose of a technical review is to maintain empirical knowledge of control operation. Therefore, the formal technical review checkpoints are always scheduled as part of initial project planning. If a meaningful anomaly or defect is identified in the course of the operation of the control system then ad hoc reviews can also be planned and scheduled to

1. evaluate impacts;
2. support functional or project management decisions;
3. carry out control design or remediation work;
4. manage the overall cyber resilience process.

Like every other type of organized activity, technical reviews begin with a formal planning process. That process underwrites and confirms resource and funding availability, and the timing and feasibility of the review schedule. The first step in the planning process is to formulate a technically capable team of testing and assurance personnel. Once the team is assembled, roles and responsibilities are assigned and the items that will be placed under review are explicitly identified and allocated to the appropriate team or team member. The review itself must be carried out by technically competent personnel. The aim is to identify all potential defects and anomalies, and document and assess their likelihood of occurrence and impact. Once that is done, the evaluator assesses whether their assigned set of controls are

1. complete and suitable for their intended use;
2. conforms to all specified performance and operational requirements;
3. conforms to regulations and standards;
4. has been correctly implemented.

Following the detailed review process, the reviewers generate and document a set of action items. That includes a specification of criteria to determine when/whether that particular action item has been closed. As with the management reviews, technical reviewers must specify

1. a unique review identifier and reviewer names;
2. review objectives for that review;
3. control(s) reviewed;
4. evidence of input to the reviews;
5. defect, vulnerability, and anomaly list;
6. resolved and unresolved anomalies;
7. itemization of action item status;
8. open versus closed action items and ownership;
9. remediation management issues.

Review Types: Inspections

Inspections are the most fundamental and frequently performed type of managerial or technical review. They are peer-based assessment activities. They detect and identify control performance or architectural anomalies. They also verify that the control and its architecture meet their specified criteria performance. They confirm that the product meets quality requirements and conforms to regulations and standards.

Outcomes of an inspection include a list of deviations from specifications, regulations, or standards. They can also collect control process development data, which can be used to improve the architectural design and the overall control inspection process.

Inspections primarily focus on some aspect of the control architecture and its individual controls. They normally comprise three to six participants and they should be led by an impartial facilitator. They must produce recommendations, but they are not problem-solving sessions. Instead, they must produce data on the control functioning, either as architecture or as an individual set within that array. They should normally examine and assess:

1. Individual and architectural control behavior
2. Control performance requirements and design documentation
3. Control validation test and user documentation
4. Control implementation and sustainment procedures

Ethically, there can be many types of participants in the review. But the personnel responsible for control design and implementation should never be the review leaders or recorders. The normal set of roles who might participate includes:

1. *Review Leader*—the single designated individual who is responsible for the conduct of process and accountable for the results
2. *Recorder*—the single person responsible for documenting a permanent record of the outcomes of the review process
3. *Reviewers and Inspectors*: the individuals comprising the review team, that might include representatives from the cyber resilience design and implementation team

A typical peer-based inspection plan would normally provide a statement of the objectives for the inspection, the control array, or overall control architecture being examined, a specific set of documented inspection procedures including all official checklists and inspection reporting forms. In addition, all potential target anomalies or issues that are being targeted by the inspection must be itemized along with all the affected specifications, regulations, and relevant performance data and anomaly categories should that are to be considered.

Inspection Procedures: When to Conduct an Inspection

The aim of an inspection is to confirm that the product is complete, correct and conforms to requisite performance criteria. Thus, inspections should be appropriately scheduled as part of the overall strategic assessment plan. There should also be a means for ad hoc scheduling of inspections as necessary. Inspections can be scheduled and held for the purposes of: control assurance management, design or implementation process support, and overall cyber resilience process management (Figure 6.5).

Since the purpose of inspections is to confirm the status of the control architecture or a control array as built, they should not be conducted until the development or implementation milestones are satisfied or the processes have been fully completed. The plan must document that the inspection process will provide

1. assurance that resourcing and funding considerations have been met;
2. the schedule was correct and feasible;
3. an inspection team comprising competent personnel was formulated;
4. all relevant roles and responsibilities were properly assigned;
5. all necessary materials and inspection items were properly distributed;
6. where appropriate decision regarding alternative courses of action or outcomes were presented.

An inspection begins with a meeting between the inspection team and the stakeholders of the affected inspection target. Obviously, if the target is the control architecture for the entire organization, that meeting would comprise upper level

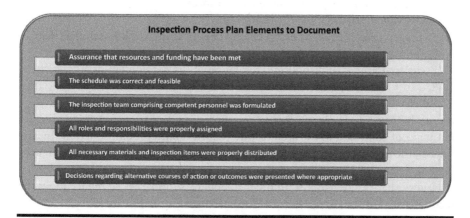

Figure 6.5 Inspection process plan elements to document.

managers and decision makers. If a particular element of that architecture or a control array is being inspected then the participants would be the managers for the functional areas where the controls apply.

The managers and review team agree on the ground rules for the inspection in that meeting. Any type of significant inspection can be intrusive. So, items like when and where the team will access the operational function, how much time will be involved, level of records produced, and participants among the regular staff and finally, reporting lines and confidentiality requirements have to be ironed out.

Once the basis is agreed on, the inspecting team meets to carry out the inspection. If a defect, vulnerability, or material weakness is identified, the inspection team documents and classifies them for face-to-face meetings with the affected stakeholders in the line organization. The product of this is a review opinion, which can choose to accept the identified anomaly with recommendations for minor rework, accept it with follow-up verification, or reinspect to verify the correctness of the rework. The point of the inspection is to produce a reliable finding. So, the final product of these discussions is a final recommendation regarding the disposition of the recommendations of the review.

Inspection Procedures: Follow-Up

The inspection process concludes with documented certification that all relevant action items have been closed. This certification is conveyed by a formal organizational document that records the

1. specific target for inspection, conceptual boundaries, and reviewers;
2. the duration of the inspection times and the scope of the documentation;
3. itemization of the particular controls inspected;
4. all evidentiary inputs to the inspection;
5. a specific description of defects, vulnerabilities, and weaknesses identified;
6. the location, description, and classification of each of those items;
7. the recommended disposition, resolution of the defect identified;
8. a cost and timing estimate of the rework required and the completion date.

As can be seen, inspections produce substantive evidence for evaluating the performance of controls and the architecture that the control is embedded in. In that respect, the organization's managers must ensure that all data that is gathered, concerning defects, vulnerabilities, and weaknesses, is stored and placed into a formal repository that can be accessed in order to update, summarize, or report about the security status of the control array or the encompassing architecture.

The defects, vulnerabilities, or weaknesses themselves must be categorized by technical type and their potential to cause harm. In addition, the cause must be

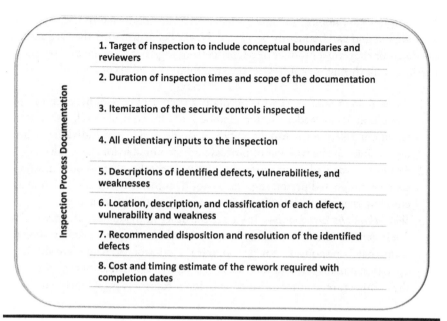

Figure 6.6   Inspection process documentation.

reported including the control elements that are either missing, extra or ambiguous, inconsistent, where improvement is desirable, not conforming to standards, or risk prone.

In addition, inspection data should be used to improve the inspection process itself. Frequently occurring defects, vulnerabilities, or weaknesses in the control array, or encompassing architecture can be used to generate inspection checklists to guide new examinations. The checklists themselves should also be evaluated for effectiveness. Resource considerations such as review preparation time and number of participants required should be considered in order to improve the efficiency of the inspection process (Figure 6.6).

Walk-Throughs

Walk-throughs are frequently used by line management to evaluate a specific set of controls while they are functioning in the operational environment. The purpose of a walk-through is to

1. detect operational problems that have not risen to the point of formal review;
2. fine-tune or improve the operational performance of the product;
3. evaluate implementations and consider the effectiveness of alternatives;
4. evaluate conformance to any relevant laws, regulations, or standards;
5. support the creative exchange of ideas and alternatives during operational use.

Walk-throughs are not as formal as reviews. They are conducted in the day-to-day environment and can frequently be ad hoc. Nevertheless, whether they are actually scheduled or they are a product of people getting together to make an operational decision, walk-throughs typically involve the following role types:

1. *Walk-through Leader*—This is normally the control development and implementation team leader and/or members of the control's staff. This is the facilitator role, and it usually entails a single designated individual who is responsible for the conduct of process and accountable for the results.
2. *Recorder*—Particularly in instances of give and take, there should always be a single designated person responsible for documenting the conversation and ensuring that any specific action items are noted for future action.
3. *Walk-through Participants*—These are the other individuals in the discussion. Their general role is to provide input to the walk-through leader and/or the team that is responsible for developing the control set. These are normally operational managers and staff who are directly responsible for the area where the architecture, control array, or individual control is being operated.

Walk-throughs are conducted in the same general fashion as reviews in that they are conducted for a specific purpose, such as to sanity test the functioning of a given control array. The first step is to identify the control or control array that is to be presented, and outline any relevant considerations in examining the walk-through target. It must be stressed that the person responsible for the creation and presentation of the initial walk-through considerations is the control stakeholder rather than the review team. That is because the stakeholder is the beneficiary of the walk-through advice.

Walk-throughs can be scheduled as part of the appropriate evaluation plan for a given control array. But generally, walk-throughs are scheduled and executed ad hoc for a particular purpose. Those purposes can include:

1. Control array or general cyber resilience management
2. To facilitate the work of control designers and implementers
3. For the purposes of performance management

Walk-throughs can be conducted on any artifact, whether it is evolving through the design process or fully implemented. Since a walk-through is informal and conducted by the stakeholder, there is no requirement for scheduling them in the strategic control development plan. However, some walk-throughs are formally scheduled to assist the designers and developers as the process rolls. The most important benefit that a formally scheduled walk-through provides is the publicity and education of the user community, line managers, and the organization with respect to the actual shape and purpose of the control architecture. Where a walk-through is scheduled, there must be a specific identification of

1. The items to be walked through including their roles and purposes;
2. the people who must attend including the designation of the stakeholder who will conduct the process;
3. the time and place that the walk-through will be conducted, that is, generally at the place where the work under review is being done;
4. any materials that will be distributed for advance knowledge of the walk-through item.

The walk-through is held as scheduled. Generally, it starts with a presentation and description of the item to be considered as well as the issues that the stakeholder wants to discuss/receive input about. If a defect, weakness, or vulnerability is identified during the walk-through then that anomaly must be documented. The means of doing that has to be specified in advance along with a description of how any anomalies arising from the element being walked through will be dealt with following the review.

It is important to keep in mind that unlike inspections walk-throughs are always under the control of the stakeholder not an impartial review team. If the aim is to not waste time, it is critical to ensure that recommended action is taken at the end of the process. The walk-through can be considered to be complete when the entire target for review has been fully presented by the designated stakeholder and the recommendations are recorded for subsequent follow-up action.

Walk-throughs will frequently produce in-process data for analysis and guidance about the effectiveness of the control and the control architecture that it is embedded in. Just as it is the case with an inspection, that data can generally be classified by:

1. Technical type and impact rating (e.g., major/minor)
2. Identification of any missing, extra, or ambiguous elements
3. Inconsistent performance or improvement desirable
4. Nonconformance with standards or risk prone

Walk-through data can be used to both improve the control architecture as well as the walk-through process itself. Frequently occurring anomalies can be documented and used to produce walk-through checklists to guide subsequent walk-throughs. Process considerations such as preparation time and number of participants can also be documented to improve the efficiency of the overall walk-through process.

Audits

Audits are generally performed by outside agency or third-party contract. Generally, those are with accounting firms. However, control architectural audits are not normally financial in nature. Instead, they provide third party certification of conformance to regulations and/or standards.

Audits are necessary when it is essential to absolutely affirm the correctness of plans, contracts, operational issues, planned procedures, formal organizational reports, and other types of formal record, such as control array and control architecture documentation.

Audits are generally sponsored and led by a person titled "Lead Auditor." That single individual is responsible for the conduct and findings of the audit. That includes performance of all administrative tasks and the overall accountability for audit opinions. Just like inspections and reviews, other administrative tasks include:

1. *Decision Maker*—the manager(s) who are being supported by the audit
2. *Lead Auditor*—the single designated individual who is responsible for the conduct of the audit process and accountable for the generation and acceptance of audit findings
3. *Auditor*—the people responsible for gathering evidence and documenting a permanent record of the audit findings
4. *Management Staff*—the management personnel in the affected organizational unit accountable for facilitating the collection, preparation, and analysis of audit results
5. *Technical Staff*—the technical personnel in the affected organizational unit accountable for the collection, preparation, and analysis of audit results
6. *Initiator*—the organization or organizational unit that has requested the audit
7. *Audited Organization*—organization or organizational unit that is the target of the audit

Initiation

Audits may be requested by outside agencies such as governmental, professional entities, the business's customers, and/or they may be self-imposed for certified assurance reasons. Because a third party is necessary to attest to organizational compliance, the auditor is almost always a third-party agency. That is, the initiator of the audit is usually not the audited organization.

The general purpose of an audit is to gather evidence that will be used to verify compliance with a well-defined and commonly understood set of laws, regulations, or requirements. The aim is to gather objective evidence about organizational control architecture or control array behavior sufficient to certify compliance with plans, regulations, and/or guidelines.

Audits are generally initiated by plan and guided by formal contract. Items that are normally stipulated in the strategic plan include resourcing and funding assurance, formal schedules for execution of the process, the process and criteria for selecting the third party to perform the audit, and the general assignment of roles and responsibilities both within the audited organization and for the auditing party. There also has to be consideration paid to ensuring the absolute integrity of the audit process and findings.

The audit itself is also planned in detail. That will always include the stipulation of roles, responsibilities, and tasks to be performed as well as the approval of the evidence gathering and statistical methodologies. At a minimum, the resulting contract for the audit will itemize the steps that will be required to

1. approve the audit plan;
2. conduct an opening meeting;
3. do the necessary materials gathering to prepare for the audit;
4. the methods that will be used to collect and examine the evidence;
5. execute the closing meeting;
6. report the preliminary conclusions;
7. report problems experienced;
8. provide recommendations for rework;
9. assure that rework has been properly performed;
10. submit the final report.

The final report will itemize and document the purpose and scope of the audit. The precise organizational boundaries of the audit target include operational functions and staff. The control procedures within that boundary have to be itemized and the audit process for those controls specified. Where there are applicable regulations and standards, those have to be stipulated along with all criteria for evaluation. The outcome of this is the audit report.

The audit report will stipulate all items that might be classified as major non-compliances as well as any potential minor or latent violations. Recommended follow-up activities for remediation are normally part of the audit report, as is their timing. The final report is concurred with by both the audited party and the auditors, and submitted as a formal record of the status of the control array or control architecture under examination.

Chapter Summary

Because they ensure the organization's designated priority assets, the control set in a cyber-resilient system must be reliably and consistently inspected and tested. The aim is to be able to say with assurance that the aggregate control set for any given cyber-resilient protection scheme is effective given the strategic aims of the organization.

Since one of the primary objectives of security control assessment is the identification of emerging weaknesses or deficiencies in the control array as implemented, organizations must conduct security control assessments during the organizational threat assessment and strategic planning phases of routine operations. The aim of these activities is to confirm the proper and correct functioning and configuration of controls as they address the known threatscape. For any organization with a formal compliance requirement, periodic control assessments of their operational systems help to satisfy the requirements of any specific legal mandates. Continuous assessments also help to better develop and implement continuous monitoring strategies for the long-term control development process.

As in most organizational projects, the process is initiated by a formal control assessment planning process. The purpose of that process is to determine the applicable scope of the overall assessment method and choose the appropriate methods as well as the target assessment objects for each control. The specification of how each general assessment effort is applied can vary from individual control components and behaviors to comprehensive evaluations of the overall performance of the control array. As we said, this is generally dictated by the requirements specified in the control model for achieving a level of detail consistent with the minimum assurance requirements of the organization.

The security control assessment plan defines the exact means that the organization will utilize to perform the specific activities of the assessment. In general, every process will attempt to verify the effectiveness of the target set of security controls. The precise actions that will be taken to accomplish that are documented in the operational steps of the plan. Those activities can include any form of standard test, audits of evidence produced by real-world control operation, anecdotal interviews of staff who are knowledgeable about system operation, and unit and integration testing of the relevant controls to determine whether they are operating as specified in the assurance case.

Assurance of that architecture is documented in the Control Architectural Assurance Plan. That plan coordinates all the diverse activities of the control assurance process. The aim is to ensure that every validation and verification process is properly related to all the relevant aspects of the project. That plan specifies the requisite assurance process and attendant resources necessary to perform all the activities and tasks needed to provide trusted cyber-resilient processes.

It must be kept in mind that the routine control tests must be customized to meet each organization's unique needs. Thus, control test planning is a critical function in control assurance. Overall control testing practices must be planned at the outset. There must be defined points where the organization pauses to thoroughly test its individual cyber resilience control requirements for their effectiveness in satisfying the organization's strategic security goals. Once the schedule for routine testing has been planned, individual control performance is evaluated to determine whether that item conforms to business requirements and the security policies and plans of the organization.

Understandably, the control testing plan brings together the results of the prior stages of cyber resilience system development, classification and risk, control design and implementation. The control testing plan is designed to continuously ensure a uniform minimum acceptable level of cyber resilience in the organization.

The first step requires the organization to delineate the specific purpose and scope of the testing process. The second consideration specifies how the testing process will be organized and conducted. It requires the organization to spell out how it will perform the control assurance.

Once the control testing plan has been developed and integrated with the other relevant plans, especially those for inspections and audits, it is time for the organization to define its standard testing procedures. That step involves formally establishing the control assurance testing activities themselves. This normally involves the definition and scheduling of tests. It also involves the institution and promotion of the necessary organizational discipline to ensure that control assurance testing is done properly and adds value to the organization.

Testing is essentially performed at several points in the control architecture development process. As we have seen, the design and implementation of controls is hierarchical. So, the control architecture evolves from an individual collection of discrete controls to a fully operational everyday architecture. Consequently, the testing process requires plans at every one of these stages.

Individual component control testing—This testing is carried out to verify that the design is properly implemented and that the individual control element or a collection of related control elements functions as specified in the operational space. The purpose of this kind of testing is to ensure that the control logic is complete and correct and that the control or controls function as intended and designed in the operational space.

Control architectural integration testing—These are a designed series of progressive tests and reviews in which the control elements are tested in the operational space as the resilience function is built. The purpose of integration testing is to ensure that the architectural design objectives are met. The aim is to confirm the correctness of the interfaces and the effectiveness of the controls as an integrated entity. This testing continues until the entire system is fully integrated.

Routine control architecture resilience testing—This is the persistent process of testing the fully integrated control architecture to verify that the cyber resilience control architecture meets its specified requirements. This testing is scheduled by plan and performed as a routine organizational function. Its purpose is to ensure that the control architecture as a comprehensive entity continues to satisfy its business overall purposes and its operational assurance objectives.

Any form of additional certification and accreditation testing—Many organizations have external requirements for security assurance. For instance, federal government organizations must meet the stipulations of the Federal Information Security Management Act. This is also true for health care organizations (HIPAA) and publicly traded corporations (Sarbanes–Oxley).

In all those instances, and many others, there is a formal testing requirement that is aimed at determining whether a system satisfies a stated set of standard criteria. The goal is to provide assurance that the organization has complied with some legal or regulatory requirement for information protection. The tests assure that the security requirements of the regulatory law or standard are met and that all the requisite controls are in place.

The specifications for performing all testing activities are specified in a formal testing plan. This plan is a formal organizational level document. It spells out the scope, approach, resources, and schedule of the testing activity. It stipulates risks and allocates resources. It explicitly states who will perform the tests, the accountable stakeholders, and how the organization will transition from one phase of testing to another. There is an explicit estimate of the number and types of test cases that will be required, the duration of the testing period, and the criteria that will indicate satisfactory completion.

Once a testing plan is completed and accepted by executive management, the tests themselves must be designed. To be effective, every testing activity needs to include an explicit statement of: (1) the required features to be tested, (2) the criteria for successfully passing the test, (3) the criteria for control performance, (4) the degree/extent of documented security performance required including the itemization of all known threats, and (5) the criteria for secure operational interface performance.

Once the test cases have been defined, the next step in the process is to develop the explicit test procedures. These procedures identify all the practical steps that will be required to demonstrate successful and effective control functioning. Each procedure is designed to satisfy the objectives of the specific test cases. Then the test procedures are executed as stipulated. The actual execution process begins at the individual control component level and proceeds up through the hierarchy of testing levels to any required certification accreditation testing.

A control can be trusted if it can be shown that it will behave in a predictable way and generate any result or output that could cause harm. The ability to trust control operation is normally termed "control reliability." That term is reinforced by a broad class of reliability assurance procedures that are undertaken to ensure that a control is safe. Conventional control assurance activities should be undertaken early in the architectural design phase of the control life cycle. The control reliability assurance process focuses on the practical behaviors of the controls and the interrelationships between the control activities. These both must be shown to be sufficient to be able to trust the control array to provide the desired protection. Consequently, the aims and purposes of the overall cyber resilience system must be completely and unambiguously understood before control reliability testing can be tackled.

Reliability metrics embrace all types of quantitative algorithm or function that can be used to obtain objective evidence of the performance of an entity, such as a control. The actual numerical value produced by a metric is called a measure. There are two general classes of metrics: management metrics, which assist in the management of the control architecture development and sustainment process and performance *metrics*, which characterize and predict the discrete behavior of the control.

Management metrics are collected and analyzed throughout the cyber resilience control development and sustainment process. These metrics are easily understandable by decision makers because they can be plotted using bar graphs and histograms or as part of statistical process control. Performance metrics are not the same as management metrics. Performance metrics are used to characterize specific characteristics or behaviors of a control, such as number of attacks prevented or successful exploitations. They are used to characterize the day-to-day behaviors of the control set. Normally, these measures lack a quantifiable basis for comparison. Therefore, they are often interpreted and used by comparison with prior plans, similar projects, or similar components within the current control array. Controls are reliable if it is impossible, or at least highly unlikely that control behavior could ever malfunction in a way that would cause an undesirable event. Examples of undesirable events include loss of information, loss of physical property, or catastrophic loss, such as total system failure.

Measurement-based assurance simply means that confidence in a given control can be assured through objective measurement. This is based on the availability of a range of standardized metrics that allow the organization to assess the performance of a given control array or architecture at a given point in time in a given situation. Over the past 20 years, there has been a growing trend toward quantitative metrics.

Reviews and audits differ from tests in that they are based on expert opinion, rather than objective evaluations of performance with respect to a defined set of testing goals. However, a review or an inspection is much easier to perform, and they can take place at any stage in the control architecture development process. They can also target both controls and the process that they operate in.

There are five types of inspections, reviews, or audits that might be performed to ensure control architectural correctness. Each of these review types involves the performance of well-defined procedures to achieve the assurance objectives. These review activities apply to either a control or its context. Management reviews ensure the correctness of management plans, schedules, requirements, control, and architectures. They support decisions about control performance, architectural effectiveness, emerging threats, corrective actions, allocation of resources, and sustainment of a requisite level of security and integrity. They directly support the management personnel who have assigned responsibility for the system. They are meant to discover and report variations from plans and/or defined procedures. They primarily focus on planning and/or corrective action artifacts.

Formal reviews are the products of an intentional strategic planning process. The aim of that process is to ensure that adequate funding and resourcing is available to carry out the assessment tasks, that the task assignments are realistic and that the overall control development schedule and inspection points are feasible, and finally, that the reporting and feedback lines are well-defined and commonly understood.

Technical reviews produce empirical evidence of control performance. They support decisions about whether the control architecture and individual controls conform to specifications of performance that are itemized in any relevant regulations or plans. The aim is to objectively determine that the control architecture has been correctly implemented or changed.

Technical reviews can be both planned and ad hoc. However, the primary purpose of a technical review is to maintain empirical knowledge of control operation. Therefore, the formal technical review checkpoints are always scheduled as part of initial project planning. If a meaningful anomaly or defect is identified in the course of the operation of the control system then ad hoc reviews can also be planed and scheduled.

The aim of an inspection is to confirm that the product is complete, correct and conforms to requisite performance criteria. Thus, inspections should be appropriately scheduled as part of the overall strategic assessment plan. There should also be a means for ad hoc scheduling of inspections as necessary. Inspections can be scheduled and held for the purposes of: control assurance management, design or implementation process support, and overall cyber resilience process management.

The inspection process concludes with documented certification that all relevant action items have been closed.

Walk-throughs are frequently used by line management to evaluate a specific set of controls while they are functioning in the operational environment. Walk-throughs are conducted in the same general fashion as reviews in that they are conducted for a specific purpose, such as to sanity test the functioning of a given control array. The first step is to identify the control or control array that is to be presented, and outline any relevant considerations in examining the walk-through target. It must be stressed that the person responsible for the creation and presentation of the initial walk-through considerations is the control stakeholder rather than the review team. That is because the stakeholder is the beneficiary of the walk-through advice.

Walk-throughs can be scheduled as part of the appropriate evaluation plan for a given control array. Walk-throughs can be conducted on any artifact, whether it is evolving through the design process or fully implemented. Since a walk-through is informal and conducted by the stakeholder, there is no requirement for scheduling them in the strategic control development plan. However, some walk-throughs are formally scheduled to assist the designers and developers as the process rolls. The most important benefit that a formally scheduled walk-through provides is the publicity and education of the user community, line managers, and the organization with respect to the actual shape and purpose of the control architecture.

It is important to keep in mind that unlike inspections walk-throughs are always under the control of the stakeholder not an impartial review team. If the aim is to not waste time, it is critical to ensure that recommended action is taken at the end of the process. The walk-through can be considered to be complete when the entire target for review has been fully presented by the designated stakeholder and the recommendations are recorded for subsequent follow-up action.

Audits are generally performed by outside agency or third-party contract. Generally, those are with accounting firms. However, control architectural audits are not normally financial in nature. Instead, they provide third party certification of conformance to regulations and/or standards.

Audits are necessary when it is essential to absolutely affirm the correctness of plans, contracts, operational issues, planned procedures, formal organizational reports, and other types of formal record, such as control array and control architecture documentation.

Audits are generally sponsored and led by a person titled "Lead Auditor." That single individual is responsible for the conduct and findings of the audit. That includes performance of all administrative tasks and the overall accountability for audit opinions. Audits may be requested by outside agencies, such as governmental, professional entities, the business's customers, and/or they may be self-imposed for certified assurance reasons.

The general purpose of an audit is to gather evidence that will be used to verify compliance with a well-defined and commonly understood set of laws, regulations, or requirements. The aim is to gather objective evidence about organizational control architecture or control array behavior sufficient to certify compliance with plans, regulations, and/or guidelines.

Audits are generally initiated by plan and guided by formal contract. Items that are normally stipulated in the strategic plan include resourcing and funding assurance, formal schedules for execution of the process, the process and criteria for selecting the third party to perform the audit, and the general assignment of roles and responsibilities both within the audited organization and for the auditing party. There also has to be consideration paid to ensuring the absolute integrity of the audit process and findings.

Keywords

Architecture: The design and implementation of an underlying framework of processes

Audit: A third party assessment that is evidence based, aimed at confirming a given status

Control design: Specification of behaviors of a protection measure or measures

Control evaluation: Formal testing or reviews of a control or control set to confirm correctness

Control performance: The operational results of control operation within a given environment

Controls: A discrete set of human or electronic behaviors, set to produce a given outcome

Critical asset: A function or object that is so central to an operation that it cannot be lost

Cyber resilience: Assurance of the survival and continued operation of critical assets

Inspection: A formal review conducted by a team comprising experts in the subject matter

Reliability: Proven capability to perform a designated purpose over time

Test: Formal hypothesis-based examination involving empirical results

Walk-through: A review that is under control of the developer rather than an inspection team

Chapter 7

Recovering the Non-Priority Assets

At the end of this chapter, the reader will understand

1. the assumptions and strategies that underlie noncritical asset recovery;
2. the functional elements of the noncritical asset recovery process;
3. how noncritical asset recovery aids in corporate resource management;
4. the assumptions associated with the noncritical asset recovery planning process;
5. the purpose and importance of recovery timing;
6. the goals and success factors in integrity points;
7. the steps involved in formulating a proper noncritical asset recovery evaluation.

You Have to Make Choices

The practical assumption that underlies the cyber resilience concept is that "you can't protect everything." So, cyber resilience spends whatever is necessary to secure critical business assets from all known threats, under every likely circumstance. The business deploys a robust collection of management processes and electronic controls sufficient to ensure that the subset of functions that are deemed essential to its continuing operation are protected beyond a reasonable level of investment by any person seeking to steal or compromise them. Thus, cyber resilience ensures that the priority assets of the organization are preserved from harm.

Logically, resources should always be allocated to ensure the survival of the vital assets of the business, while the remainder of the investment goes in ensuring the noncritical things. Noncritical assets have some value and therefore they also need a plan for protection. Practically speaking, the best argument for cyber resilience is that it concentrates resources where they will do the most good. And in that respect, the decisions that arise from the cyber resilience prioritization and design process, which we discussed in earlier chapters, must always align with the long-term benefit of the business.

Therefore, there will always be assets that cannot be fully protected. That is because the resources for assurance are either limited, or in some cases nonexistent. The assets that fall so far down the priority scale that it is not possible to justify a targeted investment to protect them are termed "noncritical assets." But since they also have some value, they need some form of protection. It can be assumed that the resources that remain after the critical assets have been fully safeguarded will be protected in two common-sense ways, traditional perimeter access controls and systematic recovery processes.

The methods and technologies for perimeter access control are well known in the domain of cybersecurity, as are routine recovery practices, such as backups. Where cyber resilience differs from cybersecurity is in the way these two functional areas specifically serve in the protection of noncritical assets. That is what this chapter is about.

Planning for Disaster

The primary difference between cybersecurity and cyber resilience is that the perimeter controls and the recovery controls are designed and deployed in a planned fashion. The planning that occurs to do that is integrated with the control deployment activity for critical assets. Cyber resilience requires the organization to spend whatever it takes to develop provably effective controls to ensure survival of those critical elements that cannot be subjected to compromise. The controls must provide verifiable assurance of core functionality as well as the various interdependencies in the enterprise's ecosystem.

However, that leaves the assets that are not organizational priorities to be protected by whatever resources are left over after the investment in absolutely ensuring the critical ones. Both critical controls and recovery controls have the same aim, which is to preserve as much of the organization's asset value as possible. But these two control concepts have different goals and purposes. In case of critical assets, the aim is to ensure the absolute confidentiality, integrity, and availability of the protection target. In case of assets to be recovered, the goal is to minimize or eliminate the impact of loss.

Even though these are two separate aims, they both have to be part of an overall strategy for cyber resilience. Therefore, both critical asset assurance and noncritical asset recovery have to be planned. The planning process for noncritical asset recovery begins when the organization reaches agreement about the assets it will unconditionally protect and the resources it will need to recover.

Recovery is a planned response that goes into effect if perimeter safeguards fail. The single goal of recovery is to ensure the rapid restoration to full operation of noncritical assets that are harmed when an attack occurs. Therefore, the goal of noncritical asset recovery planning is to lessen the impact of threats or attacks by employing well-defined plans to ensure optimum restoration of availability and integrity of assets that have some value but are not deemed critical to organizational survival.

That plan should ensure a well-defined and commonly understood process that will reestablish full functioning of the entire business operation as quickly and efficiently as possible. Thus, the plan for noncritical asset recovery must be explicit for every asset. That requires a strategic planning process that is capable of identifying, analyzing, responding to, escalating, and learning from all adverse incidents. To ensure these qualities, every asset's operating environment is examined to identify all potential failure modes. Once these modes are understood and characterized, tactics to minimize impacts on each of those individual assets are developed.

These two factors are documented in an overall noncritical asset recovery plan. The noncritical asset recovery plan dictates a detailed set of actions that will have to be taken, along with a well-defined process for the assigned roles and responsibilities to follow in order to manage and track the resolution of a failure, loss, or breach. Well-defined processes must be specified in order to ensure that all assets of value within the asset protection perimeter, as well as the operational system itself, are recovered to acceptable restoration criteria. The aim of noncritical asset recovery planning is to ensure that all system functions are fully operational within requisite parameters. This obviously requires some method or plan, which is based on input sufficient to make decisions about how to deal with any form of compromise for future recovery planning.

The plan itemizes the organization's specific approach to the prevention or minimization of damage, as well as the steps it will take to secure or recover information after the harmful event. The aim is to ensure that the organization is able to respond effectively to threats originating in its digital environment. The data obtained from systematic tests and reviews, as well as lessons learned from any compromises, will help the organization build the most effective controls for its noncritical asset recovery system, as well as conceptualize better recovery scenarios for future attacks. The primary reason why the retrospective recording of lessons learned data is so effective is that cybersecurity incidents are very diverse. However, they generally fall into eight basic attack categories as shown in Figure 7.1. They are:

1. Cyber espionage
2. Insider misuse
3. Denial-of-service (DOS) attacks
4. Crimeware
5. Web app attacks

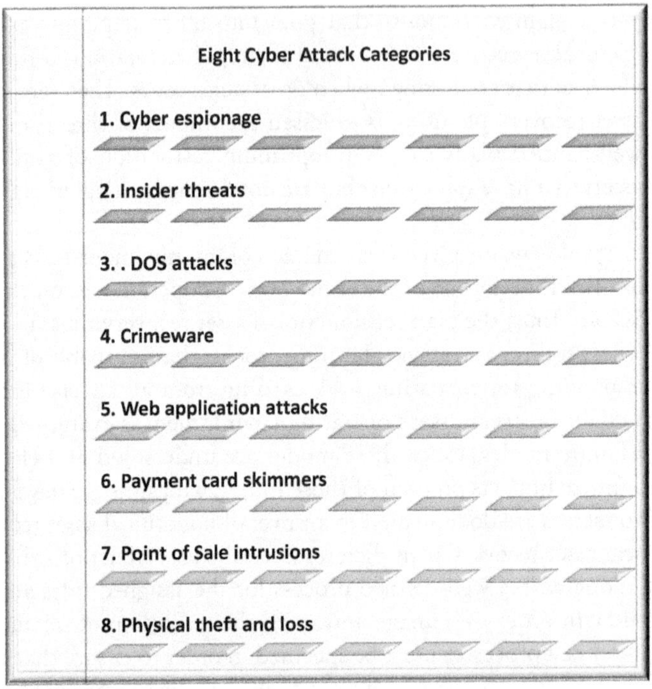

Figure 7.1 Eight cyberattack categories.

6. Payment card skimmers
7. Point of sale intrusions
8. Physical theft and loss

All these attack categories can be referenced in crafting real-world approaches to noncritical asset recovery planning for a given organization. The ability to respond dynamically to exploits based on lessons learned is also crucial to keeping costs down, because it allows the organization to "target" its preparedness planning based on what is actually happening in the threatspace. If this is done correctly, the organization will be able to achieve verifiable security at optimum cost efficiency.

Noncritical Asset Recovery Management

The goal of the noncritical asset recovery process is to design and deploy a practical business mechanism to ensure that the noncritical elements of the organization's asset base function survive in the event of an adverse event. From a business perspective, noncritical asset recovery management is primarily a planning and monitoring function.

Planning and monitoring are also basic elements of conventional management, so this process is normally a part of the organization's overall strategic planning. Nevertheless, where asset recovery management differs from the routine management process of the business is in its orientation. Most organizational planning and monitoring processes are focused on prevention, while asset recovery is focused on restoration.

The role of asset recovery is to assure the reestablishment of the company's data assets at a predefined level of integrity. The key to survival is preparation. That is the reason why strategy and planning are such essential elements of the noncritical asset management process. Noncritical asset recovery utilizes the same type of planning processes, which are used to ensure the absolute security of the organization's critical assets. The difference between critical and noncritical asset protection planning is the reconstitution requirement. Timing and integrity play a vital part in the reconstitution process. Thus, the planning for the reconstitution process must target an optimum time frame and integrity level for the reconstituted asset.

The Role of Policy and Best Practice in the Process

To ensure the optimum planned restoration, the asset recovery function adopts a well-defined and proven effective set of policies and best practices that are implemented as part of routine operating procedure. These policies and practices are specifically tailored to ensure a well-defined and disciplined response to every foreseeable harmful event.

In that respect, the noncritical asset recovery planning process utilizes three sequential steps to create a formal asset recovery response. The first of these is disaster and impact planning. Disaster and impact planning identifies and describes every foreseeable harmful contingency that might impact the organization's noncritical assets. The second activity is a logical extension of the first. The second activity deploys an optimum set of controls, which are designed to prevent or minimize the impact of every one of the contingencies that were identified in the first step in the process. The final activity is slightly different. That is restoration planning. Restoration planning is meant to ensure a controlled and disciplined recovery from a specific harmful event. The restoration planning process is different from contingency planning in that it is much more focused on developing a set of responses to a well-defined and highly specific set of occurrences. Recovery planning delineates and then maintains a detailed set of actions to be followed if an explicitly foreseen contingency occurs.

The Role of Perimeter Control

In conjunction with ensuring optimum and cost-efficient recovery processes, the organization must also consider how it will ensure perimeter control. This type of perimeter control is not to be confused with the perimeter control methods that accompany a cybersecurity approach. The difference is in the resource investment. In a cybersecurity scheme, resources invested in creating secure perimeters in a defense-in-depth model normally constitute 80% of the resource investment versus 20% for other types of security. In effect, cyber resilience reverses that percentage. Cyber resilience deploys perimeter access control measures just like an existing cybersecurity approach. It is just that the resources available to create that general perimeter control are allocated AFTER the critical assets have been proven secure by the testing and review process. That doesn't mean that basic perimeter control functions like firewalls and gate guards aren't deployed. It simply means that the resources to do that are constrained by the obligation to create verifiably secure protection for the organization's critical assets and functions.

With cyber resilience, up to 80% of the expenditure is in creating secure modules for protection of priority assets, while the 20% that is left over is devoted to creating classic perimeter access control. Those controls will generally secure the noncritical assets. Noncritical assets comprise the mundane information that organizations always keep such as public information, marketing brochures, or old sales records, anything that is considered of dubious value but kept anyway. Nevertheless, investing heavily in securing them is wasteful. That is the reason for the initial asset vetting process discussed in earlier chapters.

Thus, rather than secure perimeter control, the primary protection of noncritical assets lies in a centralized recovery function. The centralized asset recovery function comprises those principles, policies, and procedures, which ensure that an organization can recover all noncritical business functions in the event of a disaster. In that respect, the focus of asset recovery is not on preventing the disaster from occurring. Rather, asset recovery encompasses all the actions that take place to ensure business survival after the fact.

The Role of People in the Process

The aim of the noncritical asset recovery process is to ensure that all valuable but noncritical organizational resources are restored as quickly as possible to a defined state of normalcy while also making certain that those assets maintain the highest level of trustworthiness. Thus, the asset recovery function is built around a predefined set of activities and criteria that are designed to ensure that the noncritical components of the asset base are restored to a predetermined level of correctness.

Since strategies are implemented by people, the focus of most of the preparation is on dictating what an assigned set of roles will do in the face of an anticipated contingency. That includes the designation of key positions, the roles and responsibilities of those positions, the order of succession, and the communication plans that establish the chain of command. In conjunction with designating and assigning roles and responsibilities, the noncritical asset recovery planning process also makes certain that the people assigned to those roles are adequately trained and available when needed. In effect, noncritical asset recovery planning ensures that key participants sufficiently understand what to do in the face of a threat.

Also, the people responsible for implementing noncritical asset strategies must be kept current, so part of the planning process has to be devoted to ensuring that the drills and rehearsals that maintain the asset recovery management process are at the desired state of readiness and are routinely carried out.

Developing and Implementing a Noncritical Asset Recovery Strategy

The goal of the noncritical asset recovery strategy is to embed the process within the business operating parameters—i.e., the noncritical asset recovery solution must fit within the general operating concepts of the business. Accordingly, the characterization and understanding of those business parameters are vital to the success of the process, since synchronization of the recovery practices with the practices in the normal operation might be a single point of failure in the execution of the overall recovery strategy.

The aim of the noncritical asset recovery strategy is to ensure that the restoration of the target noncritical assets takes place with the highest degree of integrity in the shortest period of time. Within that scheme, this requires a strict definition of the parameters for complete and accurate restoration of noncritical assets. And thus, a clear understanding of the noncritical asset timing and restoration requirements, for that particular business application, might be the single starting point for determining all the rest of the process and practice requirements for that particular situation. As we saw, effective asset recovery is based on the following related but independent factors:

1. The time it takes to initiate a restoration process
2. The time it takes to recover noncritical assets to their desired operational status
3. The time it takes to ensure that the data has been restored to its proper level of integrity

Each of these things is independent in the sense that there are a multitude of unique factors that determine their successful execution. However, there are five questions, shown below and in Figure 7.2, that need to be answered for every one of these factors if you want to ensure that the plan embodies the right set of practical elements:

1. What are the business functions that will require a noncritical asset recovery?
2. What is the business impact of each noncritical asset?
3. What are the risks associated with noncritical asset restoration timing (RT)?

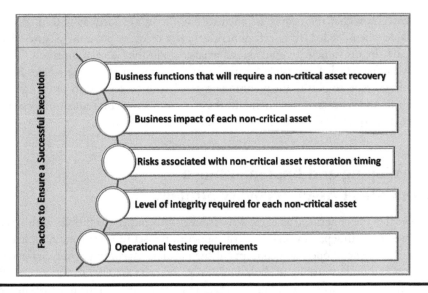

Figure 7.2 Factors to ensure a successful execution.

4. What is the level of integrity required for each noncritical asset?
5. What are the operational testing requirements?

Developing and Maintaining an Effective Response

The organization is responsible for developing and maintaining strategic perspective in the development of its asset recovery management plans. As such, those plans are developed through a formal strategic planning process. The asset recovery management plan dictates, in very specific terms, the exact procedures required to restore operation of the information processing function. Each procedure has the specific assignment of staff and resources associated with it. That includes both timing and cost information for every resource. These procedures are documented down to the level of the exact steps that will be taken to prevent or minimize damage.

One of the broader aims of the asset recovery management plan is to ensure operational resilience through timely recoverability of the assets that have not been locked down in the critical asset protection architecture. In that respect, the asset recovery management plan must demonstrate that the organization is positioned to maximize the potential confidentiality, integrity, and availability of all noncritical assets. Practically speaking, the preparedness plan lays out the critical path that the organization must follow to ensure timely restoration of any asset that has been harmed, compromised, or removed from access. In that respect, it is very important to specify the criteria for evaluating the integrity of that organization's constituent no-critical data.

That kind of detailed preparedness planning is driven by the same threat modeling, risk assessment, and threat analysis processes that we discussed in Chapters 3 and 4. Hence, many of the activities that support the creation of a data security management process also support the specific formulation and maintenance of the enterprise's preparedness plan. Thus, the preparedness strategy should be developed in conjunction with the development of the organization's critical asset security architecture, since the overall value preservation aims of the two functions are the same.

The preparedness plan should continue to evolve with the threat environment and the realities of the business climate. As such, preparedness planning must always be cognizant of the status of three basic components of the overall information processing function, the data—including all critical information baselines, the physical facilities and equipment, and the capabilities, roles, and responsibilities of the information processing function's personnel. That final category includes all operational staff, not just critical staff.

The Preparedness Plan

An essential component of the preparedness planning process is the identification of business contingencies to address. Every foreseeable operating contingency has to be recognized and properly characterized for the plan to be successful. Then management must develop a strategy and attendant set of actions to ensure that each of these contingencies has a concrete and specifically designated response associated with it.

Since the aim of the adversary is to compromise the confidentiality, integrity, and availability of an organization's data assets, much of the asset recovery process is built around developing secure backup and storage approaches. That implies the need for extensive contingency planning.

The development of contingency plans is typically based on threat assessment. Threat assessments identify all forms of threat. However, only those contingencies that require a formal response are actually incorporated in the planning. Every one of the formal responses is then documented in a preparedness plan. Essentially, the specific actions that are stated in the preparedness plan represent the steps that the organization plans to take to address every contingency of significance.

Operationally, the exact steps involved in ensuring that each threat is properly addressed are detailed in the preparedness plan. The purpose of the plan is to document the strategic approach that the organization will follow to respond to each of its priority contingencies. The aim of the plan is to ensure a substantive across-the-board reply that will offer the best trade-off between the harm that a given threat might cause and the cost of putting the requisite actions in place to mitigate it.

The key principle in preparedness planning is the issue of foreseeability. In simple terms, foreseeability just means that a threat is sufficiently well anticipated and understood to allow the organization to take specific planned actions to mitigate the harm. Those planning assumptions are developed as a response by a threat scenario. The aim of the threat scenario is to come up with an overall set of thoroughly well-thought-through plans to cover every reasonable consequence that is identified for the situation. Each scenario is updated as additional knowledge is gained about the probability of occurrence and harm of a given threat. These scenarios are then prioritized. That prioritization is typically based on the assumed likelihood and impact that would result from each threat.

The ability to identify and prioritize threats relies on the type of threat modeling and estimation methodologies that were discussed in Chapter 3. Various estimation models and tools can be utilized to support the threat identification and response planning process. Those methods and tools might include such common practices as threat models or probability of occurrence and net-present-value estimates of economic impact. Then, once the organization fully understands the consequences of each threat scenario, it prepares and subsequently maintains a practical asset recovery management response.

Risk Assessment and Preparedness Planning

The information needed to support a preparedness plan is obtained by means of a conventional risk assessment process. The role that risk assessment plays in asset recovery planning is to help develop the timing of the restoration. It is important to be able to prioritize the actions to be taken for the restoration since that is the way decision makers assign resources.

Those decisions usually require a little bit of fortune-telling on the organization's part. That is because the threat scenarios have not yet happened, and indeed might never occur. Consequently, the preparedness plan typically prioritizes responses by category or type of harmful event. The aim is to address any potential eventuality that might fall within a given class of event with a single response. For instance, whether the loss of information is due to a major hacking exploit or a simple human error, the outcome of losing a given piece of noncritical information can be foreseen, functionally described, and a mitigation planned. Therefore, the preparedness plan can specify the steps to react to any form of asset loss including everything from insider theft to total loss of the equipment that processes the asset.

Developing an Effective Preparedness Plan

Cyber resilience preparedness plans seek to ensure that the enterprise will continue to function in the face of a wide range of potential negative occurrences. The tangible outcome of the recovery planning process is a cyber resilience preparedness plan or CRPP. The CRPP specifies the explicit steps that will be followed to ensure optimum recovery from a specifically stipulated occurrence. Because the impacts of a specific occurrence are known and can be characterized in advance, this allows the organization to have an optimized response in place prior to the occurrence.

The goal of preparedness planning is to minimize the potential harm to the confidentiality, integrity, or availability of an information asset, and the associated interruption of business should an incident occur. The types of recovery that must be accommodated will vary based on the specific requirements of the situation. Thus, one of the initial practical requirements in doing recovery planning is to understand the timing requirements.

Timing plays a large part in shaping the organization's preparedness planning. That is, strategic decisions such as when a particular response will be utilized, and what the desired outcomes are of that particular response, are dictated by the length of time for which the system and its information are expected to be unavailable. Extreme decisions, like the decision to change the business model due to information loss, typically apply only if the disruption will be for an extended period of time. However, in most cases, preparedness planning decisions are made based strictly on the current business case.

Consequently, a proper preparedness plan is built on a detailed understanding of the impact that the loss of the asset will have on the overall business processes. In that respect then, preparedness planning involves two things. The first is a set of validated assumptions about the circumstances and the events surrounding a given asset. The second is the resultant strategy for ensuring asset recovery and subsequent reintegration into the business process.

Step One: Assumptions

Preparedness planning assumptions are based on an understanding of the threat environment. This is a dynamic situation in the sense that the threat picture is constantly changing, and so the assumptions have to be periodically updated. As new risks emerge, assumptions have to be made about their impacts on critical business functions. This could include the timing and extent of the threat as well as the areas of potential harm.

Step Two: Priorities and Strategy

The second element is more conventional, that is, the development of a noncritical asset recovery strategy. Organizations formulate their noncritical asset recovery strategies out of their particular business philosophies and the values of their assets. Therefore, in that respect, there is no such thing as a uniform approach. Each business approaches the challenges of prioritizing and developing noncritical asset recovery responses in its own way. The only given in this process is that the strategy that is eventually adopted and the philosophy that drives it must be uniformly understood and accepted throughout the entire organization.

The latter condition is an absolute must because there cannot be confusion about how to respond to a disaster when it occurs. As such, it is important that the organization adopts and then communicates a single common noncritical asset recovery strategy. Furthermore, to make the approach acceptable to employees, the strategy should originate from and align with the business philosophy of the organization.

In addition to a commonly understood approach, it is important to know exactly when, where, and how the plan applies. For instance, an effective noncritical asset recovery plan would not be designed to kick in at the exact point where the breach occurred and the data were lost. The organization would instead build in precautionary drills, assessments, and remediation procedures.

So, the questions that need to be answered are: When will the noncritical asset recovery process start and when will it stop? As an example, in case of a breach, it would be very hazardous to have the noncritical asset recovery plan terminate the minute the intruder exited the system because intruders can always come back. Therefore, the noncritical asset recovery strategy would also include the remediation and rework requirements needed to seal the breach. On the other hand though, it would be extremely resource inefficient to be still following the emergency procedures stipulated in your noncritical asset recovery plan the year after a breach occurred. So, it is important to embed a set of unambiguous criteria that will dictate when the noncritical asset recovery process will begin and when it will terminate.

The Recovery Process

Recovery has to be approached from both long-term and short-term perspectives. In the long term, there has to be a well-thought-out process in place to ensure successful recovery from all breaches and losses of information that fall within its scope. Nevertheless, in order to guarantee that the long-term recovery process will continue to remain effective, it is important that short-term processes exist to continuously classify threats, assess their potential danger, and devise effective responses with respect to a given threat.

So, recovery planning also has to document ways to maintain a balanced and realistic understanding of only those noncritical assets that could cause a loss of business value. In order to do that, the organization should perform a regularly scheduled update of the threat assessments and recovery scenarios that drive the recovery process.

In the short term, asset recovery managers have to have in-hand a precise set of steps, which they will follow to respond to a particular recovery challenge. These are called recovery scenarios. The contingencies that drive each recovery scenario have to be precisely specified in advance, and then justified and synchronized with the procedures that will be followed in the event that a given eventuality should occur. The specification is always embodied in an explicit statement of the steps to be taken, expressed in very clear and unambiguous language. These steps must be referenced to each individual circumstance.

Thus, recovery planning is built around a specification of the work to be done in the event of a given eventuality. This specification is primarily aimed at dictating the exact roles and responsibilities of the people in the process and the specific actions that need to be taken in the face of a given threat.

This is a behavior-centric specification that requires the development of an attendant set of highly focused motivational and educational processes. Those processes have to ensure that the managers and other stakeholders responsible for implementing and maintaining the plan are properly up-to-speed in what they have to do.

Finally, recovery plans have to be tested. This is an absolute requirement for the formal implementation of a recovery process. It also requires that the planners both refine as well as operationally test their assumptions on an ongoing basis, with the aim of proving that the specific approaches embodied in the recovery plan for that item effectively address the requirements of each specific scenario.

Documenting a Recovery Plan

The cyber resilience recovery planning process documents a formal set of asset recovery actions. The tangible outcome of the planning process is a detailed prescription of the set of real-world steps that will be taken in response to any event that falls under a given set of assumptions. Asset recovery management plans always involve two things. The first are the specific statements that document and justify why particular contingencies have been chosen while other potentially foreseen contingencies were not selected. The second is a complete and detailed specification of the strategies that will be used to ensure that the necessary asset recovery steps are taken in case of each contingency.

The rationale behind the first item is that there will always be more threats than can be feasibly mitigated. However, threats are always time-sensitive, since new threats can arise at any time and old threats can become more dangerous. Therefore, the assumptions that guided the assignment of priorities have to be made explicit in order to steer the future evolution of the recovery response.

These assumptions are always based strictly on the organization's threat picture at the current point in time. Assumptions have to be periodically tested and updated because that picture is constantly changing. One important side benefit of maintaining the assumptions in alignment with the known threat picture is that it legally documents the organization's due diligence in maintaining its asset recovery management response should a breach or loss occur.

As a new threat is identified, a set of assumptions has to be developed about its impact on critical business functions. The assumptions will normally be based on the likelihood and impact of the threat as well as the areas of the business that might be subject to harm. Nevertheless, given the uncertain nature of most threats, there is always something of uncertainty factor in the process. However, the outcome of the process of thinking through and documenting priorities is always the same. All the assumptions that underlie the organization's asset recovery management strategy have to be maintained current and as relevant to the threat situation as possible.

The second element of the design process is much more directly focused, that is, the development of the actual asset recovery management strategy. Organizations devise asset recovery management strategies out of their own particular business philosophies. Given the relationship between business goals and the strategy that is adopted, there is no such thing as a uniform method for development. Every organization approaches the challenges of ensuring asset recovery in its own unique way. The only given in the process is that the strategy that is eventually implemented as well as the philosophy that drives it must be uniformly understood and commonly accepted throughout the entire organization.

That strategy must be well understood and commonly accepted. This is an absolute requirement for success because there cannot be disagreement about how to respond to an adverse event if it occurs. As such, it is important that the organization both implement and then fully communicate a well-defined asset recovery management plan for every noncritical asset of interest. Moreover, in order to make it acceptable to all the employees, this strategy must be fully sponsored by executive management with concrete accountabilities for performance built into the process.

Because adverse events do not usually arrive in neat packages, it is also important to embed a set of unambiguous criteria into the overall asset recovery management strategy that precisely and unambiguously dictates when the plan will be activated and when it will terminate. The precise criteria for activating the asset recovery management plan have to be understood by every essential person in the asset recovery management process, and regular drills have to be undertaken in order to ensure that those people understand exactly when, where, and how their roles are involved.

Elements of the Asset Recovery Plan

Although the list of potential threats might be long, the plans that address them always encompass the same three elements: (1) threat description and impact/likelihood classification, (2) threat recovery response deployment and recovery method and communication processes, and (3) escalation and reassessment procedures.

Threat impact classification requires the organization to understand and describe the practical implications of the threat. For instance, knowing that a worm-based DOS attack is likely going to occur does not really address the consequences that might result. The questions that must be thought through in the recovery plan are the specific impacts of having a given DOS attack actually happen. Once all contingencies have been identified, classified, and their likelihood assessed, a formal set of steps to mitigate them can be prepared.

Because a threat never plays itself out in exactly the same way in every situation, the formal response to threat is generally based on the execution of a preplanned set of responses involving a set of assigned roles. That is, the preparedness plan designates the right people to be deployed and the responsibilities they will have in the case of a given threat.

This creates the maximum flexibility when it comes to executing the desired response. For instance, since a DOS attack can come from many sources, it is impossible to dictate in detail the exact way to react to a distributed denial-of-service (DDOS). However, it is sufficient to know who is specifically responsible for performing the planned actions for mitigating it, for instance, contacting the upline services to install the filtering. Likewise, a proper plan will designate which services to contact as well as their detailed contact information.

Finally, because cybersecurity attacks involve a considerable helping of the unknown, there has to be a defined set of escalation and reevaluation procedures. A well-defined set of escalation procedures is needed in order to ensure an adequate response should events continue to escalate or not meet the planning assumptions. Thus, the plan has to include a specific and well-defined organizational communication and decision-making path for reevaluating and reclassifying the threat situation.

Identification and Prioritization of Protected Functions

The logical first step in creating a specific asset recovery management plan for the organization's noncritical assets of value is to identify and characterize all the things that will fall under that plan. In essence, the business follows the same process that it employed to identify and prioritize its priority assets. In fact, in general, the noncritical asset identification process can be performed as part of the overall identification and classification of the priority assets.

First, the functions that are not essential to ensuring the continued operation of the business are identified. A lot of routine information is generated in the business day. This information is a by-product of the business processes. But it does not rise to the confidentiality, integrity, and availability requirements necessary to make it "critical" or "priority." In essence, this information can be reconstituted from backups, or even lost without any direct, meaningful impact on the general business purpose.

The common-sense question to guide the priority asset identification process is whether the loss or temporary unavailability of that information could lead to some significant loss of value in the overall business operation. If the answer is "no" then the asset is classified as "non-critical" and the recovery function applies.

Once all the noncritical information has been identified and characterized, which can represent the bulk of the information and information processing assets of the business, a priority is assigned to each noncritical item. That priority is based on the process' assumed value to the overall business purpose.

Even though the assets are noncritical, asset recovery management planning requires a priority listing of all functions to be recovered. That is because the next step in the process will develop an asset recovery management solution for each asset that is considered worthwhile to recover by a systematic solution. The difference between cyber resilience versus cybersecurity is that the organization is willing to sacrifice some of its nonessential data in order to ensure that its essential data is fully protected. However, some nonessential data is more important than other data. So, a second dividing line has to be drawn between investment in the assets that will be risked but recovered and those that are deemed not sufficiently worthwhile to protect. This is a simple resource commitment estimate, which attempts to balance the cost of the protection with the value of the protection target.

The recovery of noncritical assets will always require a commitment of resources. So, the pragmatic aim of this part of the process is to ensure that the noncritical assets with the highest value get the resources that they need, both in terms of backup solutions and timing. This is the same sort of sliding scale estimate that is done with priority assets; the most valuable noncritical assets will receive the greatest investments.

Once a list of noncritical assets is prepared, the resource commitments associated with sustaining those assets must be designated. This is an important step because the organization has committed whatever is required to protect its critical assets. So, in essence, what's left over is what sustains and protects the noncritical assets. The remaining resource availabilities have to be well defined in order to make the planning effective. Since that might not be much in terms of resources, the decisions about the noncritical assets that need to be preserved have to be wisely and judiciously made.

The organization first identifies the internal resources that will be utilized in the asset recovery process, such as specialized staff or internal backup capabilities. Then it identifies the external resource contributions, such as third-party services or alternate sites.

A resource specification at this level of detail lets the organization identify potential shortfalls in either the resources or capabilities to support the proposed plan. These shortfall areas should be itemized and then justified to the overall strategy for cyber resilience. Shortfalls indicate where the organization does not have the resources or where it does not have the capability to meet a given recovery objective. That justification can lead to a more cost-effective approach to the overall process.

It is important to identify shortfall areas, because those are the potential points of failure in the recovery plan. Knowing where the shortfall areas are benefits the long-term sustainment of the process. That is because that knowledge tells managers where adjustments to their cyber resilience strategy are necessary.

Accordingly, the process of adjusting the plan to fit within available resources is an extremely critical element of the asset recovery management planning process. It should then be possible to create a feasible cyber resilience recovery process once all the asset recovery management's resource requirements have been appropriately justified to the business priorities of the organization.

Executing the Asset Recovery Process

The goal of the asset recovery process is to ensure the resilience of noncritical business assets whenever there is a loss or breach. The operational plan for shifting the organization from normal operations to asset recovery management operations has to be as detailed as the steps in the recovery process itself. Thus, the decision-making process with respect to when and how to execute the noncritical asset recovery plan becomes a plan in-and-of itself.

Logically then, there are two basic ways that asset recovery can be approached: planned and unplanned actions. Planned actions work well when the attack or breach can be foreseen, such as a power failure or a well-understood type of attack. Having a set plan in place also allows the organization to make last minute adjustments to their approach and can foster a more orderly, lower stress response.

The transition to planned asset recovery operations should be as automated as possible. This automation can prevent loss of business functions, business connections, and business transactions by shifting the operation instantaneously to an alternative source. Nonetheless, whether the shift is accommodated automatically or as a result of advanced planning and meetings, the same approach is used to shift functional business systems from normal operations to planned asset recovery functions. The plan becomes active at the point where an emergency has been identified.

However, some threats to noncritical asset are unanticipated. Consequently, there is no advance planning for the response. In that case, however, a formal protocol for the characterization and reporting of the novel threat's behavior must exist. This is in effect a plan for how to develop a disciplined response to a previously unanticipated emergency. The plan must identify who is to be certainly involved in the emergency response, such as the response manager. It must also provide a standard evaluation and planning protocol for any form of generic threat, and it must define formal notification and reporting lines for whatever response is determined. The aim is to provide a disciplined and well-organized systematic mechanism for dealing with unexpected threats or violations. The creation of such a plan is as essential to the noncritical asset recovery process as is the formal planning for anticipated events.

Two Essential Factors

The aim of asset recovery management is to ensure the preservation of the enterprise's noncritical information assets. In that respect, the approach that is taken to the process of preservation is dictated by two fundamental factors. The first is the timing of the recovery. The second is the level of reliability that has to be guaranteed for the recovered data. Given those two factors, the design process requires the organization to establish two fundamental aims. These are restoration time (RT) and the integrity level.

In simple terms, recovery time specifies the maximum acceptable down time for the information processing operation integrity level specifying the minimum level of trustworthiness that is required for the restored data. Ensuring the timely access to, and reliability of, the organization's data is the whole point of the asset recovery management process. Therefore, these two fundamental building blocks will define the precise shape of any particular organization's asset recovery management process.

The period of time that it takes to recover the asset is a business decision. As a result, the choice about where to set the recovery time is based on the information processing demands of the business. In terms of the three fundamental principles of information assurance, confidentiality, integrity, and availability, recovery time represents the "availability" principle. The recovery period defined by each recovery time goal is a discrete measure of time, which defines the maximum operationally acceptable interval for which a specific information system or information processing function can be unavailable before unacceptable harm occurs.

The trustworthiness of the data is dictated by the backup point (BP). Since the data recovery process generally depends on identifying the last point where the data was known to be correct, trustworthiness is usually defined by the last BP. Because any data entered on the system after the last backup cannot be trusted, the data recovery BP is typically set at the shortest feasible interval. That is because data that is created or modified outside of that interval has to be considered lost or in need of painstaking restoration procedures to ensure its validity.

The BP is a different concept altogether from timing. Because integrity is an essential element of information assurance, every asset recovery plan has to specify a minimum acceptable level of certifiable surety for the recovered data. In simple terms, the BP just indicates how much data loss the organization is willing to accept in light of its business goals. Obviously, business considerations like value and investment have a lot to do with setting the BP in each particular instance.

There is a direct trade-off between the value of data and the cost of backing it up. Thus, the desired recovery point objective and the investment to ensure that that objective is met are generally traded off against each other. The aim is to arrive at an optimum balance between the investment in assurance and the value of each item of data. In that respect then, the practical case might be made for investing whatever it takes in real-time backups to ensure the reliability of the data that is vital to the organization's survival. On the other hand, the value of data that is nonessential might not justify a backup of more than once a week.

Elements of the Backup/Restoration Solution

A lot of the actual implementation process for a restoration strategy involves ensuring the most effective backup mechanism possible. In operational terms, restoration from backups requires the organization to create and then maintain the ability to switch its information processing function from an impacted primary system to a preplanned and matching alternative system. The ideal situation would be to conduct fully redundant information processing operations sufficient to ensure the availability of one of the systems without loss of data integrity.

However, except in rare cases, the value of noncritical data might not be sufficiently great enough to justify the cost of operating mirror systems, even though this would ensure the highest level of integrity and availability of all data, both critical and noncritical. Consequently, the simple rule-of-thumb is that the level of backup redundancy must be directly aligned to the actual value of the data. The idea that the perceived value or criticality of the data determines the level of redundancy introduces the notion of alternative data recovery approaches.

The pragmatic aim of the noncritical asset recovery and restoration process is to be able to restore the affected data in the shortest and most realistic practical time. Consequently, in order to understand how alternative data recovery works, it is necessary to understand the concepts of recovery time and recovery point.

The estimate of recovery time is derived by determining the Maximum-Tolerable-Loss time (MTL) for the noncritical data of a given function. MTL is an adaptation of the old-fashioned data processing term, maximum-tolerable-downtime. It applies in case of asset recovery because the organization has to know how much the loss of a piece of noncritical information will affect the overall functioning of the business process.

Ideally, the aim of an asset recovery management plan is to ensure the integrity and availability of all meaningful noncritical items of data. The best means for doing that is to back that information up to an alternative storage location, one that is not subject to direct attack.

By definition, noncritical information is not crucial to the actual survival of the business. Therefore, the backup estimates are generally based on when the information might be needed. That involves three types of estimates:

1. RT—e.g., when is this needed?
2. Network Recovery Timing (NRT)—e.g., if transmitted, when is it required?
3. BP—e.g., what is the timing of the backup?

RT is a business concept. It is a timing goal that is based on the process itself. An item of noncritical data might be needed as soon as a critical operation is restored; payroll records for instance. However, some items might be only useful as an archival record, or never be needed again. This all has to be sorted out as part of the decisions that are made about the timing of the noncritical data backup and recovery process.

NRT is an adjunct to RT. This term specifies the greatest amount of time for which a network that accesses noncritical assets can be out of service. This estimate is driven by the same business requirements as RT. That is because in modern business organizations users access information through networks. So, the same practical timing considerations apply, as they do for the noncritical information assets they transmit. In that respect then, network recovery requirements have to be considered along with RT estimates in restoration and recovery planning.

BP timing is a different concept altogether. It is concerned with the timing of the backup, which in turn impacts information integrity, as opposed to its availability. The BP timing estimate identifies the point in time that the data must be certifiably correct. Data created or modified outside of the parameters set by the timing estimate is not reliable by definition.

Therefore, among other things, the BP will determine what items of noncritical information can and cannot be trusted. Consequently, every asset recovery management plan must specify the points in the everyday operation when a particular noncritical asset needs to be backed up. As may be the case, archival information may only be backed up once.

Critical information is a separate concern, which is part of the priority item aspect of the cyber resilience process. However, many items of noncritical data are updated with the critical items in the business process. The timing of the backups dictates the presumed integrity level of the information. In case of noncritical items, which are of some operational value to the organization, these BPs must occur at fairly short intervals. Therefore, since noncritical items are also useful, their assurance also requires an estimate of the appropriate BP.

Ideally, the backup would have happened at the exact point where the breach or harm occurred. But in order to maintain that rigorous level of assurance, it would be necessary to adopt an automated solution that maintains a mirror backup during the everyday business operation. This might be economically or technically infeasible. More often than not, organizations must make conscious trade-offs about the most effective BP within their particular business context. For instance, it is the necessity to assure trustworthiness that drives the decisions about whether to adopt a cloud solution versus one that is more "hands-on," such as routine business process-based backups to a local bulk storage utility.

RT, NRT, and BPs are not mutually exclusive concepts. But they are different in that they imply a different set of practical decisions as well as a different set of resource requirements. For instance, because of the cost of having workers idle, a highly automated automobile assembly plant will want to set the shortest RT possible for their assembly lines. However, they could possibly afford to lose some of the production data without a great deal of financial impact. Therefore, they might be willing to trade-off the general timing of the backed-up data recoveries to meet that requirement.

In contrast, if a bank's systems are to remain credible, they have to be able to absolutely ensure that backups of customer transaction data are maintained to the exact point of failure, even if the actual time it takes to bring the system back on line is extended. Therefore, the recovery time and the NRT might be extended in order to ensure that the critical integrity issues associated with the backup are met.

Making the Noncritical Asset Recovery Process Real

The timing of the recovery is assigned based on the previously determined criticality of the component in the overall business operation. The integrity requirements are assigned based on the level of trustworthiness that the restored data must achieve in order to be useful to the business. As we said earlier, because of resource constraints, timing and integrity typically trade-off against each other on a sliding scale. That is, where greater integrity is required a longer recovery time can be tolerated. Whatever the relationship between timing and integrity, however it is a rule of good planning that every important one of the noncritical data items that are protected under the cyber resilience recovery process must have an explicit timing and recovery statement associated with it.

Because the practical concern lies with the ability to provide sufficient resources to achieve a stated level of timing and integrity, it is essential that a precise estimate of the resources required to accomplish these proposed levels accompany the plan. Typically, that statement involves an estimation of the number and types of personnel who will be required to do the work, along with each person's duties, where they will be housed, and any training requirements. A detailed resource estimate always accompanies final noncritical asset recovery plan.

Specification of Recovery Actions

Operationally, recovery actions have to be specified down to the level of the personnel who will perform each action, the technologies that will be used in that process, and how alternative sites will be accessed and utilized if needed. The specification of procedures has to address all the potentially significant operational and environmental factors, including technology, personnel limitations, and geographic factors. It should also incorporate any specific plans for migration of the noncritical items to another platform, or the utilization of contracted or outsourced services during the time a function is unavailable. Finally, because organizations change constantly, there is the implicit requirement to update the asset recovery management plans on a regular, if not continuous, basis.

Identification and Documentation of the Solution

All the actions needed to satisfy the specified availability and integrity requirements for each noncritical item contained in the cyber resilience recovery management plan are recorded on an individualized statement of work (SOW). This SOW is a specification of the practices that will be employed to both sustain the recovery process and in case of ensuing harm should a known threat occur. This document should describe the requisite procedures and the accompanying resources that will be deployed in case of any loss of noncritical assets.

The SOW documents all of the organization's assumptions about noncritical asset recovery. It has to be based on a careful analysis of the various business circumstances and environmental conditions for a given organization. This SOW itemizes the explicit procedures to be carried out, down to the level of prescribed behaviors for each individual or process, for that particular function. That includes the identification of requisite personnel, work area, equipment, and supply or service capability, along with any procedures for making up a shortfall in resources.

The SOW is meant to provide the clearest possible guidance on the actual steps that will be taken to address every contingency that can be foreseen. Finally, in order to ensure the continuing applicability of the recommendations in the SOW, it is important to perform regular audit evaluations of the steps specified in the plan in order to determine whether they are all still applicable and correct.

Ensuring that Everybody Knows What to Do

The formal documentation and acceptance by upper management effectively completes the noncritical asset recovery management planning. The set of recommendations that constitute the noncritical asset recovery management plan itemizes the organization's approach to recovery management. Detailed procedures are specified to ensure the availability and integrity of the noncritical but protected assets.

Assuming that they have been well thought-out, these procedures should assure proper asset recovery for the organization's noncritical but necessary functions and services. Nonetheless, the people who are responsible for implementing and managing the plan itself should be fully aware of what is required of them. As such, one final step remains.

The business must be able to guarantee that all the participants in the process clearly understand their assigned roles and responsibilities. Operationally that assurance amounts to ensuring that the people who are responsible for performing the actions that have been detailed in the noncritical asset recovery plan are capable of carrying out their assignments.

The actual identification of the people who will participate in the plan should have taken place at the time that the noncritical functions were documented. Normally, the managers who are responsible for implementing and overseeing the asset recovery management function are given a list of the key people who might be designated for the recovery work for each function. These lists serve as the basis for the assignment of tasks, as well as the estimation of the need for any other types of resources.

Once the assignment of roles and responsibilities is complete, the organization needs to plan and then execute a focused, top-to-bottom training program. The goal of that program is to bring the people who have been assigned roles in preserving noncritical assets up to the requisite level of capability. The outcome of this training and education program is that everybody who is essential to the success of the noncritical asset recovery management plan is fully capable and aware of his or her duties.

Finally, from a human factor perspective, it is also critical that all the levels of management, from senior managers down to team and project leaders, are given explicit awareness of the role and function of the noncritical asset recovery management process. That is because the organization's management must actively support the process. Asset recovery is a resource intensive process which, like insurance, only really makes its value evident after the fact. Therefore, it is hard to maintain the necessary degree of management discipline to ensure that the process remains effective without strong endorsement and support from the top.

Operational Considerations: Trading Off the Two Factors

RT and recovery BP are two separate concepts, which are in some respects mutually exclusive of each other. That is, if you want to get the system back to running in the shortest possible time, it is usually harder to ensure the trustworthiness of the data. So, the optimum recovery point versus recovery time also has to be decided. Organizations where computers drive the business, such as automated manufacturing plants, generally favor the shortest possible recovery time.

On the other hand, in an institution where the credibility of the data is more critical to the organization, a brokerage house or a research laboratory for instance, the integrity of every byte of data has to be ensured. As a result, an optimum trade-off point has to be decided between how soon the system is restored versus how trustworthy the data will be in designing an asset recovery management solution.

It is that need to find this balance that drives the practical decisions about the shape of the recovery process. Ideally, the point where the information was backed up would be positioned exactly at the point where the loss occurred, meaning there would be no lost information. But in order to maintain such a high level of assurance, it would be necessary to run a continuous backup operation. Since that would entail the cost of maintaining a real-time backup process at an alternative location, this approach would probably be economically infeasible for most organizations.

Because businesses perform a wide range of functions, some more-or-less critical than others, there is both a general and a specific aspect to setting the RT and BPs. From the management standpoint, in order to implement a practical asset recovery response, it is essential to have a single RT and recovery point for the business.

Such a single fixed point of restoration is necessary in order to allow managers to make strategic decisions about the resource and business aspects of its operation.

Nevertheless, because most business operations are comprised of diverse functions, it would be unrealistic to think that every aspect of the business can be brought back into operation at the exact same recovery time and BP. Nevertheless, it is still possible to determine recovery times and recovery points for each of the individual, noncritical processes of the business. Then once these are known, it should be possible to aggregate the average of these into a practicable schedule of RTs and BPs for the overall business.

Evaluating the Noncritical Asset Recovery Process

The goal of asset recovery is to restore the noncritical asset base as close to the occurrence of an adverse event as is feasible. Given that aim, however, the asset recovery management process also needs to fit within the specific operating parameters of the business. That practical requirement implies the necessity to perform a comprehensive and detailed business analysis in order to ensure that the asset recovery function and the day-to-day operation of the business are properly aligned.

Consequently, the asset recovery management process must ensure that the actions that will be taken to recover and restore noncritical asset are compatible with the needs of the business and its underlying information processing operation. To ensure that compatibility, the design process needs to make certain that a set of practical requirements have been met.

First, the cyber resilience recovery development process should ensure that every one of the relevant, affected business functions are identified. Second, the design process must ensure that the relative degree of contribution of each of these functions, with respect to the overall operation, is known. Third, the specific hazards that threaten each business function and their likelihood of occurrence and impacts should be described. Then, the practical restoration criteria for noncritical assets affecting each function can be determined from this analysis. Finally, an aggregate average recovery time and recovery point for the business are set.

In practice, the organization passes through five logical stages in order to create a substantive noncritical asset recovery function. In the first stage in that process, ALL the noncritical functions that are considered vital to the business are identified and prioritized. The prioritization is based on the overall value of each function to the business and the likelihood and impacts of any threats involved. Next, the specific noncritical asset recovery solution is prepared for each function. That solution sets the most effective recovery time and recovery point for each function. An overall noncritical asset recovery management plan is then developed using the aggregate of RTs and backup requirements. Finally, the organization establishes an overall asset recovery management plan and puts the practices in that plan into place.

It probably goes without saying that it is important that the asset recovery management plan is feasible. Yet, there is no standard rule for how to ensure feasibility. However, there IS one absolute requirement. That is that all noncritical functions are assigned a specific and explicit recovery time and integrity level. Then, the backup and timing for each function must have a specific set of backup and recovery procedures associated with each of them. These must be geared to the timing requirements. Moreover, it must be possible to validate by direct observation that each of those procedures are present in the normal business operation. In addition, it must be possible to demonstrate that those procedures are functioning properly.

Factors that Affect the Noncritical Asset Assurance Operation

The global assumptions and their resultant formal processes must be properly assessed in order to correctly implement a noncritical asset recovery management solution. In addition to the practical implementation and the continuous evaluation of the operational system's performance, evaluation also helps refine the solutions that are in place as the organization evolves.

At a minimum, the evaluation mandate obligates every organization to fully and completely identify the risks to its noncritical assets and the impacts that their failures might have on the overall business operation. There are three kinds of conventional analysis activities that are associated with the development of noncritical asset recovery management plans. Those are business impact analysis, risk analysis, and the redundancy determination.

Impact Assessment

Business impact analyses are carried out to determine the impacts that a potential asset loss might have on a business function. These analyses must consider a range of eventualities when evaluating the solution to a given problem. Those could include everything from the inability of the organization to properly document a business transaction, all the way up to factors such as financial or even legal and regulatory exposures. Based on the outcomes, such an assessment might promote a noncritical asset to critical status as the organization evolves.

There are always marginally critical assets in any business. Therefore, considerations that are part of such a business impact study might include such factors as impact on revenues and expenditures should the asset be compromised, as well as any application recovery influences that might leverage the greater long-term importance of a given asset; there even needs to be consideration of possible fines or other kinds of regulatory penalties if noncritical information were lost.

Risk Evaluation

The risk analysis evaluates the risks represented by any threat to a noncritical function. That includes the underlying resources that support those functions. Once a threat has been identified, risk analysis seeks to estimate the likelihood that any given set of noncritical assets will suffer meaningful harm or disruption. This assessment determines the degree of impact and likelihood for each individual item. Once that risk level is understood for a given noncritical asset or collection of assets, it is usually possible to make strategic decisions with respect to whether the risk is worth addressing and, if so, how to approach addressing it. It is this assessment that supports decisions about the availability and integrity points of the solution.

As we said, the resources required to establish and maintain immediate or short-term availability are dictated by the criticality of the information being protected and the degree of assurance required. Business impact analyses are normally carried out to determine what the consequences are of the loss of the data in some particular aspect of the business operation.

As we also said, these analyses normally involve the use of threat scenarios to drive the actual estimations. Those scenarios can range as wildly as the imagination of the planners and the limitations of the business case. However, resource-efficient threat estimations have to focus on a set of likely threats. For instance, planners might consider such pedestrian issues as whether the organization could sustain operations in the face of the loss of its advertising brochures. But they can also range all the way up to significant disasters like the complete loss of financial records or the theft of proprietary information.

Given the wide range of possible considerations, the actual criteria that drive the business assessments are rarely strictly technical. For instance, it is perfectly appropriate in a business impact analysis to consider the effect of such happenings as a reduction of the funds to support preparedness or an unanticipated increase in the cost of maintaining the response at a given level of desirability. It is also appropriate to consider legal and regulatory hazards such as the potential for fines or other kinds of penalties if critical data were not available to support a Sarbanes–Oxley or a Health Insurance Portability and Accountability Act (HIPAA) audit.

Robustness

A large component of noncritical asset recovery planning is the estimation of the robustness of a given item or a collection of noncritical assets. Robustness attempts to describe the survivability of a particular asset or asset collection in the instance of a concerted effort to harm or subvert it. Obviously, robustness factors are a lot more extensive and rigorous for critical assets. However, there also has to be a consideration of just how robust noncritical items are in the face of concerted attacks.

In case of critical assets, there is almost no room for failure and the protection solution has to be deployed accordingly. In case of noncritical assets, the worst outcome from a loss or breach might be nothing more than inconvenience. Consequently, the costs of failure are much less dramatic. However, since all assets that have some value need some form of protection, the relative outcomes of their compromise have to be factored into the noncritical asset recovery planning.

The redundancy estimate determines the resource allocations for what amounts to backup assurance. There is normally some form of general backup protection for the organization's assets. It might be nothing more than spinning the information out to tape. But that investment represents a minimum level of assurance. The problem is that assets all require a varying degree of protection. So, the organization has to get a good grip on what that variance implies if it wants to spend its security dollars wisely.

At its core, the general aim of noncritical asset recovery is to guarantee that the availability of all the organization's information assets is continuously assured and that the integrity of those assets is continuously maintained. Because both availability and integrity require the electronic elements of the system to operate in an uninterrupted fashion, this implies a need for some form of redundancy. The normal approach to ensuring this is as old as data processing itself, well-defined and systematic backup operations.

Ensuring the Continuing Effectiveness of the Response

The evaluation of the effectiveness of any organizational process is a necessary feature for establishing the necessary accountabilities and discipline. Without the disciplined execution of the process, it is difficult to ensure the long-term sustainability of the solution. In case of the noncritical asset recovery management process, periodic evaluations are aimed at assessing the effectiveness of the procedures specified in both the critical and noncritical asset recovery plans.

Regular evaluation of the effectiveness of the noncritical asset recovery management process ensures that the activities that take place within that process are justifiable over time. The main purpose of such an assessment is to identify operational problems, that is, situations where a requisite backup procedure is either missing or ineffective. If problems are identified, the role of the evaluators is to notify the stakeholders who have been made accountable for bringing a deficient process back into alignment.

The asset recovery management evaluation process, as utilized in the day-to-day business world, verifies the continuing effectiveness of a particular planned approach. Evaluation ensures that this approach will leverage a timely recovery. In addition, occasionally, the evaluation processes might be called upon by third parties in order to verify compliance with some aspect of a contract or regulation.

In practice, the evaluation process assesses the operational execution of the planned noncritical asset recovery management procedures in the light of a set of specific scenarios or assumptions about a potential threat. In that respect, the aim of the evaluation process is to certify the ability of that procedure to satisfy its planning assumptions. The evaluation always assesses three things: the currency and effectiveness of the policies or assumptions that underlie the procedure, the ability of the designated participants to carry out the requisite procedures, and finally the capability of the relevant managers to oversee the overall process.

In all three instances, the outcome of the evaluation process ought to be explicit evidence that the right set of policies are in place and that the corresponding correct set of practices are being followed to achieve the stated objectives of a given plan. Moreover, in addition to assuring the continuing effectiveness of each of the requisite activities in the plan, the noncritical asset operational evaluation process also helps the organization refine its response over time. At a minimum, the evaluation process involves routinely assessing the two dominant focuses of the asset recovery management process, business impact and threat response.

Finally, at this level of detail, it takes a lot of time and effort to understand and then be able to reliably ensure the practical management needs of the process. In addition, because organizations change constantly, it is necessary to perform this assessment on a regular, if not continuous, basis. Because such an assessment process can be time and resource intensive, the commitment to sustain the process must be understood.

Nonetheless, depending on factors such as the size of the organization and the level of real assurance expected, as well as the practical considerations of resources available, it is both possible and acceptable to limit the noncritical elements that could be covered by the plan. There is one absolute requirement however: Those activities designated as within the scope of the noncritical asset recovery strategy must be appropriately considered, and it must be possible to validate that they have been addressed by direct observation. If that confirmation is not possible, the scope of the strategy must be adjusted to exclude assets that cannot be addressed by a planned solution.

The strategy can be confirmed as feasibly scoped, the way the resources will be provided must be developed, agreed to, and cross-referenced to the resource estimate. Depending on the situation, the resources to achieve a given level of noncritical asset recovery can come from a number of places, involving both internal and external sources. The organization should first identify the internal resources that will be provided to support the plan and then any external sources, such as those of contractors. Once this analysis is done, the organization should be able to identify any potential shortfalls in either resources or capabilities. These shortfall areas should be itemized and cross-referenced to the existing noncritical asset recovery strategy. It is very important to identify and deal with the shortfall areas because there are potential failure points. Therefore, this part of the process is an extremely critical element of ensuring a well-executed noncritical asset recovery strategy.

This resource information is recorded as an explicit set of commitments to address any foreseeable problems. That includes a suitable set of recommendations for how any existing shortfalls will be addressed. This statement should also stipulate the organization's assumptions about the various circumstances and contingencies that will bring about that investment, that is meant to provide the clearest possible resourcing guidance for every foreseeable contingency.

Finally, in order to ensure the continuing applicability of these recommendations, it is good practice to perform regular audits of the operational activities in order to determine whether the resourcing assumptions still apply. Audits are also important to determine whether the noncritical asset recovery strategy has been adequately documented. There are commercial services that will perform routine third-party assessments of the capability of an organization's asset recovery management plan. These might be utilized as well.

Chapter Summary

Practically speaking, the best argument for cyber resilience is that it concentrates resources where they will make the most difference. And in that respect, the decisions that come out of the cyber resilience prioritization and design process must align with the business's long-term investment strategy. As a result, the resources are always allocated to ensure the survival of the long-term strategic assets of the business, while the remainder goes in ensuring the noncritical things.

Logically then, there will be assets where the resources for assurance will be limited or nonexistent. But since they have some value, those assets also need assurance. It can be assumed that the resources, that remain after the critical assets have been protected, will go into two common-sense areas, traditional perimeter access controls and systematic recovery processes.

The methods and technologies for basic perimeter access control are well known in the domain of cybersecurity, as have routine recovery practices, such as backups. Where cyber resilience differs from cybersecurity is in the way these two knowledge areas serve the protection of noncritical assets. That is what this chapter was about.

The single goal of recovery is to ensure the rapid restoration to full operation of noncritical assets that have been harmed when an attack occurs. Therefore, the goal of the recovery planning is to lessen the impact of disruptive events by employing well-defined plans to minimize loss and/or unavailability of the organizational assets that have some value but are not deemed critical to organizational survival.

The plan for recovery must be explicit for every asset. That requires an operational plan capable of identifying, analyzing, responding to, escalating, and learning from all adverse incidents. To ensure this, every asset's operating environment is considered to identify all potential failure modes. Once these modes are understood and characterized, tactics to minimize impacts on each of those individual assets are developed.

These two factors are documented in a preparedness plan. The preparedness plan dictates a detailed set of actions that will be taken along with a well-defined process for assigning roles and responsibilities, and managing and tracking resolutions should a disaster occur. The plan itemizes the organization's specific approach to the prevention or minimization of damage as well as the steps it will take to secure or recover information after a harmful event. That plan should ensure a well-defined and commonly understood process that will reestablish full functioning of the entire business operation as quickly and efficiently as possible.

The aim of the recovery process is to ensure that any noncritical organizational data assets are restored as quickly as possible to their pre-disaster state while also making certain that those data assets maintain the highest level of integrity. Thus, the asset recovery function is built around a predefined set of activities and criteria that are designed to return the noncritical components of the asset base back to a predetermined level of correctness. The key to survival is preparation. That is the reason why strategy and planning are such essential elements of the asset management process.

The organization is responsible for establishing and maintaining a strategic perspective in the development of its asset recovery management plans. As such, those plans are developed through a formal strategic planning process. The asset recovery management plan dictates, in very specific terms, the exact procedures required to restore operation of the information processing function. Each procedure has the specific assignment of staff and resources associated with it. That includes both timing and cost information for every resource. These procedures are documented down to the level of the exact steps that will be taken to prevent or minimize damage.

Cyber resilience recovery response plans seek to ensure that the enterprise will continue to function in the face of a wide range of potential negative occurrences. The tangible outcome of the recovery planning process is a cyber resilience recovery plan (CRRP). The CRP specifies the explicit steps that will be followed to ensure optimum recovery from a specifically stipulated occurrence. Because the impacts of a specific occurrence are known and can be characterized in advance, this allows the organization to have an optimized response in place prior to the occurrence.

The goal of a recovery planning is to minimize the potential harm to the confidentiality, integrity, or availability of an information asset, and the associated interruption of business should an incident occur. The types of recovery that must be accommodated will vary based on the specific requirements of the situation. Thus, one of the initial practical requirements is doing recovery planning.

Recovery planning has to be approached from both long-term and short-term perspectives. In the long term, there has to be a well-thought-out process in place to ensure successful recovery from all breaches and losses of information that fall within its scope. Nevertheless, in order to guarantee that the long-term recovery process will continue to remain effective, it is important that short-term processes exist to classify threats, assess their potential danger, and maintain the effectiveness of the response with respect to a given threat.

Although the list of potential threats might be long, the plans that address them always encompass the same three elements: (1) threat description and impact/likelihood classification, (2) threat recovery response deployment and recovery method and communication processes, and (3) escalation and reassessment procedures.

The logical first step in creating a specific asset recovery management plan is to identify and characterize all the things that will fall under that plan. In essence then, the business follows the same process that it employed to identify its priority assets. In fact, the noncritical asset identification process can be incorporated into the one that is adopted for priority assets. First, the functions that are not essential to ensuring the continued operation of the business are identified. A lot of routine information is generated in the business day. This information is a by-product of the business processes. But it does not rise to the confidentiality, integrity, and availability requirements necessary to make it "critical" or "priority."

Asset recovery management planning requires a priority listing of all functions to be recovered. That is because the next step in the process will develop an asset recovery management solution for each asset that is considered worthwhile to recover by a systematic solution. Besides the commitment to recovering critical items, the recovery of noncritical assets will also require a commitment of resources. So, the pragmatic aim of this step in the cyber resilience recovery development process is to ensure that the noncritical assets with the highest value get the resources that they need, both in terms of backup solutions and timing.

Once a list of noncritical assets is prepared, the resource commitments associated with sustaining those assets must be designated. This is an important step because the organization has committed whatever is required to protect its critical assets. So, in essence what's left over is what sustains and protects the noncritical assets. The remaining resource availabilities have to be well defined in order to make the planning effective.

The goal of asset recovery is to restore the noncritical asset base as close to the occurrence of an adverse event as is feasible. Given that aim, however, the asset recovery management process also needs to fit within the specific operating parameters of the business. That practical requirement implies the necessity to perform a comprehensive and detailed business analysis in order to ensure that the asset recovery function and the day-to-day operation of the business are properly aligned.

Consequently, the design of the asset recovery management process must ensure that the actions that will be taken to recover and restore noncritical asset are compatible with the needs of the business and its underlying information processing operation. To ensure that compatibility, the design process needs to make certain that five practical requirements have been met.

In practice, five logical stages are followed to create a substantive asset recovery management plan. In the first stage in that process, all the functions that are vital to the business are identified and prioritized. The prioritization is based on the overall criticality of each function to the business and the likelihood and impacts of any threats involved. Next, the specific asset recovery management solution is prepared. That solution sets the most effective recovery time and recovery point for each function. The overall asset recovery management plan is then developed using the aggregate of recovery times and recovery points. Finally, the organization establishes an overall asset recovery management plan and then puts practices in place to ensure that the asset recovery solution remains effective.

The timing of the recovery is assigned based on the previously determined criticality of the component in the overall business process. The integrity requirements are assigned based on the level of integrity standard that the restored data must satisfy in order to be useful to the business. Because of resource constraints, timing and integrity typically trade-off against each other on a sliding scale. That is, where greater integrity is required a longer recovery time can be tolerated. Whatever the relationship between timing and integrity, it is a rule of good planning that every important one of the noncritical data items that are protected under the cyber resilience recovery process must have an explicit timing and recovery statement associated with it.

All the actions needed to satisfy the specified availability and integrity requirements for each noncritical item contained in the cyber resilience recovery management plan are recorded on an individualized SOW. This SOW is a specification of the practices that will be employed to both sustain the recovery process and in case of ensuing harm should a known threat occur. This document should describe the requisite procedures and the accompanying resources that will be deployed in case of any loss of noncritical assets.

The goal of the asset recovery process is to ensure the resilience of noncritical business assets whenever there is a loss or breach. The operational plan for shifting the organization from normal operations to asset recovery management operations has to be as detailed as the steps in the recovery process itself. Thus, the decision-making process with respect to when and how to execute the noncritical asset recovery plan becomes a plan in-and-of itself.

Logically then, there are two basic ways that asset recovery can be approached: planned and unplanned actions. Planned actions work well when the attack or breach can be foreseen, such as a power failure or a well-understood type of attack. Having a set plan in place also allows the organization to make last minute adjustments to their approach and can foster a more orderly, lower stress response.

The global assumptions and resultant formal processes must be properly assessed in order to correctly implement a noncritical asset recovery management solution. In addition to the practical implementation and the continuous evaluation of the operational system's performance, evaluation also helps refine the solutions that are in place as the organization evolves.

At a minimum, the evaluation mandate obligates every organization to fully and completely identify the risks to its noncritical assets and the impacts that their failure might have on the overall business operation. There are three kinds of conventional analysis activities that are associated with the development of non-critical asset recovery management plans. Those are business impact analysis, risk analysis, and the redundancy determination.

A lot of the actual implementation process for a restoration strategy involves ensuring the most effective backup mechanism possible. In operational terms, restoration from backups requires the organization to create and then maintain the ability to switch its information processing function from an impacted primary system to a preplanned and matching alternative system. The ideal situation would be to conduct fully redundant information processing operations sufficient to ensure the availability of one of the systems without loss of data integrity.

Regular evaluation of the effectiveness of the noncritical asset recovery management process ensures that the activities that take place within that process are justifiable over time. The main purpose of such an assessment is to identify operational problems, that is, situations where a requisite backup procedure is either missing or ineffective. If problems are identified, the role of the evaluators is to notify the stakeholders who have been made accountable for bringing a deficient process back into alignment.

In practice, the evaluation process assesses the operational execution of the planned noncritical asset recovery management procedures in the light of a set of specific scenarios or assumptions about a potential threat. In that respect, the aim of the evaluation process is to certify the ability of that procedure to satisfy its planning assumptions. The evaluation always assesses three things: the currency and effectiveness of the policies or assumptions that underlie the procedure, the ability of the designated participants to carry out the requisite procedures, and finally the capability of the relevant managers to oversee the overall process.

Keywords

Business impact analysis: Routine evaluation of the business environment in order to ensure continuing relevance of the noncritical asset recovery plan

Contingency planning: Planning based on identifying all likely scenarios

Escalation procedures: Procedures that allow the organization to respond in a disciplined fashion to an unforeseen event

Impact classification: Prioritization of a harmful event based on its impact or harm

Noncritical asset recovery assumptions: Assumptions about the legitimate scenarios and actual events that might take place for a given noncritical asset recovery plan

Noncritical asset backup point: Minimum acceptable level of assured integrity for a dataset—another term for the integrity principle

Noncritical asset recovery planning: Planning to ensure that noncritical functions and data of value to the organization are sustained in the event of disaster

Noncritical asset recovery strategy: Specific broad-scale plan to ensure the successful recovery of noncritical assets

Noncritical asset recovery timing: Maximum period of time that the data can be unavailable without harming the business—another term for the availability principle

Noncritical asset recovery plan: The specific steps the organization will take to recover its noncritical information or information processing function following a foreseen event

Noncritical asset recovery scenario: Assumptions about the course of a given scenario or threat action embedded in the noncritical asset recovery plan

Operational assessment: Assessment of the effectiveness of a noncritical data asset recovery plan which is performed in the operational setting

Optimum trade-off: Finding an optimum relationship between RT and BP based on economic factors

Preparedness plan: The organization's specific strategy to respond to foreseen disasters

Prioritization: In planning the order that a threat will be addressed based on the potential harm it will cause

Protection strategies: The specific strategies and procedures incorporated in the preparedness plan

Restoration: The process of ensuring that data is restored to a desired level of integrity following a disaster

Robustness: The ability of a process or system to withstand attack—the resiliency principle

Threat analysis: Continuous operational analysis of the threat environment in order to identify any changes that will have to be planned

Chapter 8

Ensuring a Continuously Cyber-Resilient Organization

At the end of this chapter, the reader will understand

1. how to create and sustain a long-term cyber resilience program;
2. the standard elements of sustainment architecture;
3. tailoring out functional elements of noncritical asset recovery from standards;
4. the assumptions associated with sustainment architecture;
5. the purposes and importance of documentation and archiving;
6. the goals and success factors of the audit process;
7. the steps involved in ensuring long-term sustainment.

It All Starts with Infrastructure

This chapter presents and discusses the detailed principles that underlie the development of dependable, large-scale organizational infrastructures. Those principles are rooted in strategic process definition and planning. The purpose of both of those comprehensive processes is to establish continuing and consistent day-to-day functional cyber resilience operations.

Specifically, this chapter will detail a practical organizational process for developing and maintaining a stable, complete, and sustainable cyber resilience response. The goal of cyber resilience is to ease the impact of disruptive events on the information asset base. This assurance is guaranteed by utilizing well-defined organizational control processes to protect critical and noncritical information assets and services. In real-world terms, the necessary trust is underwritten by the creation and sustainment of an operational control infrastructure for cyber resilience.

The goal of this formal structure is to foster and maintain a complete and consistent environment for the conduct of the everyday cyber resilience process. The infrastructure makes the abstract elements of cyber resilience real and tangible to the people in the organization. In order to do this, the infrastructure formally documents and embodies all the relevant organization-level strategic and operational concerns that might be required to successfully ensure critical asset protection. In addition to protection, the infrastructure will also embody and sustain an effective and reliable noncritical asset recovery mechanism.

Infrastructure development also embodies the practical steps that are required to ensure the continuous improvement of the conventional cyber resilience process. An organization's cyber resilience infrastructure ensures suitable execution of the everyday practices needed to ensure protection of critical assets and recovery of noncritical ones.

Organizations maintain strategic governance mechanisms as a means of ensuring that the fundamental processes and practices that comprise the cyber resilience infrastructure have been fully and correctly aligned to the particular organizational purpose. Governance also ensures that those processes are executed in such a way that they ensure the long-term, effective achievement of the overall business goal of full and complete protection of their information assets. The infrastructure makes that abstract notion real and substantive, in the sense that it oversees and regulates the day-to-day operational activities that perform the cyber resilience process.

Logically, the specification of those activities must be both detailed for any given organizational application as well as strategically viable as part of the overall conduct of the business case. These divergent goals are satisfied by a well-defined and formal array of deliberately interrelated operational processes that are capable of identifying, analyzing, responding to, escalating, and learning from all adverse events.

Embedding the Cyber Resilience Process in Day-to-Day Business

In order to be effective, the processes that comprise a cyber-resilient infrastructure have to be performed as part of the overall day-to-day operation of the business. Thus, cyber resilience must be embedded in the conventional business as an everyday operational process. That is the case because every form of behavior that is designed to mitigate risk must be continuously functioning in order to ensure against the exploitation of vulnerabilities. The real-world behaviors that comprise that continuous protection are generically termed persistent or organizational "controls."

Persistent control is an important concept in architecture because it represents the practical behaviors or actions to be performed. The control universe is by necessity diverse and complex. That is because the threats and attacks they are tasked to address are often intricate and multifaceted. More importantly, they are not always computer based. Exploits like social engineering and insider theft center on the human factor rather than the machine. That implies that it must be possible for companies to oversee and manage an increasingly complex set of day-to-day practices that must be reliably executed by every member of the organization. Given the layers of complexity associated with something as abstract as virtual work, this is a particularly difficult task.

Therefore, the creation of a reliable cyber resilience architecture requires a wide variety of precisely targeted behavioral processes and automated controls. The real-world realization of these controls must be

1. planned based on intimate knowledge of the threat environment;
2. fully integrated to produce a synergistic result;
3. economically feasible.

Most of today's global business is carried out in distributed, multilayered, multivendor, and even multicultural environments. All that work must be satisfactorily coordinated and controlled in order for the company's stakeholders to trust the outcomes. That coordination and control are often very difficult to achieve given today's complicated business challenges.

Therefore, oversight and control of complex work are built around the assurance of a common and stable infrastructure of management controls. A control infrastructure helps the organization guarantee uniform and consistent outcomes from all of the many operational components of a multifaceted corporate environment. That common point of control is embodied in the everyday control architecture of the organization.

Given the inherent complexity in control formulation, the only way to ensure a systematic architectural solution is through a single rational development strategy. The strategy must ensure the trustworthy, long-term cyber resilience capability of the organization. And, it must address all likely threats to all items of critical and noncritical assets. A solution such as that must be systematically designed to merge all requisite behavioral and technical controls into a single, fully coordinated operational process across the organization.

The architecture must be implemented as a substantive element of organizational functioning and then evolved as a way to meet the changing needs of the environment. The generic term for this type of large-scale infrastructure management process is "governance." Because it becomes harder to ensure reliable outcomes as business organizations grow and diversify, a rationally planned governance architecture of realistically targeted controls is important for any form of successful vulnerability management.

Throughout this text, we have approached the topic of cyber resilience as if it is a strategic governance function. The purpose of adopting that stance is to avoid a piecemeal solution. Whether they are formally documented or not, every enterprise is managed by a set of commonly accepted practices. In most corporations, those practices represent an individual unit's or manager's understanding of the proper way to carry out a specific task, and they tend to embed themselves in the organization over time.

Typically, every organization's cybersecurity practice has evolved this way. That is, organizations have developed solutions bit by bit as the situation arose rather than as an across-the-board strategic governance process. The problem is that a piecemeal protection architecture doesn't work in a world of persistent threats. That is because the operational security response will not fully embody or address all the concerns for the asset base.

The alternative is to formally define, design, and install a comprehensive array of management controls, which are aimed specifically at optimizing the effectiveness of the security function. This is done as a coordinated entity. As with any complex deployment, these controls can only be substantiated through a rational and explicit planning process. The practical management term that is used to describe a process of designing and deploying a specifically targeted universal infrastructure response is "security architecture." That term effectively describes the strategic approach and purposes of the cyber resilience development process. Moreover, it is this strategic architectural concept that drives the deployment of the types of controls we have been discussing in this text.

Security Architecture

We need to spend some time explaining the general principles embodied in security architecture. The term "security architecture" was coined to describe the strategic function that underwrites proper due diligence in the assurance of the organization's information assets. Security architecture consciously builds an intentional structure of rational elements and interorganizational relationships that will sufficiently ensure the business' information assets.

A security architecture approach centers on the creation of a comprehensive organizational control system and security culture rather than building separate individualized security solutions for each problem. Essentially, the business defines a coherent framework that embodies all the necessary roles, their associated behaviors and practices, and the strategic policies necessary to achieve a single coherent, organization-wide cyber-resilient solution. There are five processes that underwrite the development of a security architectural framework.

Scope

The first of these principles is scope. In essence, scope addresses protection issues in the light of resource constraints. In the worst case, that might involve leaving a nonessential or low risk resource outside the boundaries of the protection scheme.

Proper scoping entails the execution of a deliberate process to establish the boundaries of the solution. In some respects, it is the most critical step of all, since the extent of the territory that must be covered by the security solution will dictate the form and extent of the rest of the process, and in the real world that extent implies a conscious balancing between need and the resources to meet that need.

Accordingly, the underlying issue that scoping addresses is: "How does the organization get optimum assurance out of its resource investment?" In day-to-day practice, this means that it must be possible to make an informed and intelligent decision about the level of risk that can be accepted within the resources available. Obviously, anything can be secured if enough money is thrown at it. But no organization has the money to effectively put a cop on every street corner. So, a deliberate process has to be undertaken that balances deployment of the assurance response against the likelihood and material consequences of the threat. Factors that might enter into this process include considerations such as: What is the level of criticality for each information asset and what is the degree of assurance required for each?

During cyber resilience prioritization (Chapter 4) that determination is captured on a ten-point asset classification rating scale that ranges from "not needed" on one end of the scale all the way up to "the business would close without this" on the other end. This can be used to support the tough decisions that will have to be made at the intersection between the decision to protect or leave out the defense.

Because technology evolves, the process of establishing a cyber resilience scheme is always dynamic. For decision makers, proper scoping maximizes resource utilization and provides the foundation for the rest of the process. In case of scope, this means that the protection modules are subject to ongoing refinement. That refinement is based on information feedback from other activities, particularly the risk assessment.

In essence then, although the first step is to define the boundaries of the modules that will protect the critical assets, these are not fixed. They are always subject to change as the realities of the environmental circumstance dictate. However, if that decision is made based on precise knowledge about an established perimeter, this knowledge is obtained by formal assessment.

Standard Risk Assessment

Assessment is the second essential element of the cyber resilience infrastructure development process. That is because assessment is a primary player in any form of risk-based security system development. The assessment function ensures that the organization fully understands the risks inherent in its security environment and the associated set of requirements.

Since risks don't come with convenient labels, criteria are required to assess risks. Those criteria are normally embodied in the form of a requisite best-practice reference model, which documents all the necessary controls of a standard governance model. In that sense, the detailed recommendations of a governance model comprise the basis for risk assessment.

That best-practice model should be capable of optimizing the security governance process for that organization and represent expert consensus. Thus, most commonly accepted best-practice standards are documented as a logical structure within a domain and process framework. The model itself should allow an organization to evaluate the business risks and assign control behaviors to a common set of best-practice functions. By definition, a control objective is a precise statement of the desired result or purpose to be achieved by implementing a given control procedure for a particular activity. The standard that is utilized to make that decision should provide comprehensive criteria to make decisions about the overall completeness and correctness of a managerial control system for a given set of organizational assets.

In that respect, there are several reference models that can be selected. However, the Information System Audit and Control Association's (ISACA) control objectives for information and related technologies (COBIT) standard, the International Standards Association's (ISO) ISO 27000, and the National Institute of Standards and Technology's (NIST) NIST 800-53 are arguably the most frequently referenced models.

Any one of these frameworks can be used to judge whether a security infrastructure and its constituent activities are complete and correct. Whatever the source, the reference framework should specify and populate a well-defined and detailed organizational best-practice architecture that allows the organization to develop its own explicit control policies, practices, and procedures. In their practical application, COBIT and ISO 27000 are primarily oriented toward conventional business. NIST 800-53 satisfies the requirements of the Federal Information Security Management Act (FISMA) and therefore it is almost exclusively used in government.

All these frameworks start from one simple and pragmatic assumption: The operation must be managed by means of a set of logically related everyday behaviors, which taken as a whole constitute a complete set of security best practices for an organization. ISO 27000 has 14 process areas, the COBIT Framework defines four general process areas and a set of 34 high-level control objectives, and NIST 800-53 embodies 17 process areas. These are assumed to describe and embody all aspects of information and information technology (IT) functioning. By satisfying the specified requirements of each area, the manager can ensure that a capable IT control system is in place.

The actual control system is tailored top-down to any desired level of detail for any given organizational application. A standard tailoring approach is particularly important in the creation of security infrastructures, since the implementation of the actual security system is always different in its particulars. A standardized security control infrastructure, which is defined at a commonly accepted, correct level of detail provides the communal policy and procedure reference points that are essential for the reliable definition of a practical security solution for a given application. Or put more simply, because a standard framework is stable, it can be trusted as the basis for making the rational trade-offs and adjustments that are always implicitly necessary to formulate a specific cyber resilience infrastructure solution.

Why is a well-defined and commonly accepted standard-based approach effective? First, it provides a continuous point of reference to reflect and accommodate the full range of difference in the security needs of the pertinent organizations. The ability to deal effectively with a wide range of potential applications is an essential quality in any standard-based security design process. That is because the threat environment is so wide ranging and diverse. Persistent threats can be identified at any time and from any source. So, the security response has to be creative and multifaceted. Since every potential application of the standard is different, the implementation process has to accommodate that range of difference.

However, since it is clear that there is a need for flexibility in the tailoring of controls from a standard, there are also good reasons for standardization. Standardizing the controls on a model of best practice ensures the time-tested effectiveness of the solution. Plus, with standard control areas, each implementation of the specific standard solution can improve the process through lessons learned. Standardization provides a standardized basis for measurement and metrics. And finally, in the practical world of business, it is naive to suggest that unique security solution should be developed ground up for every organizational application. So, some sort of standardized approach to the problem is also an absolute prerequisite for effective use of security resources.

In order to be at their most effective, security infrastructure development processes should have both maximum flexibility and standard structure, which might seem like a contradiction. The simple resolution is to create the architecture of the solution top-down from the highest possible level of concept. The largest and most comprehensive view of the solution can then be used as a general classification structure within which effective controls for each of the specific security areas can be addressed. Using this approach, a specific cyber-resilient control architecture can be constructed for any given project at any given level of definition inside the model best-practice framework.

However, the classification structure itself must always be consistent. That is, the overall control recommendations of the general framework model must provide the uniform best-practice elements and structural relationships that will allow a specific cyber-resilient control architecture to be refined to any desired level of detail for every project. In theory, using the approach, an optimum process can be defined for a given project.

The outcome of the specific tailoring of controls to satisfy each of the general security categories of the standard should be an operational process model, that embodies best practice, and which specifically ensures cyber resilience for a given organizational application. Practically speaking, a well-defined and documented practical infrastructure of controls that is developed from one of the control frameworks mentioned above is of considerable value to the business. That is because that documentation makes the overall cyber-resilient control architecture and its attendant procedures tangible, and communicates them to the employees of the organization.

The behavior of any form of control system has to be documented. Documentation is an absolute ongoing requirement. In many respects, the documentation of control outputs is the absolute bedrock on which oversight and management rest. So, it would not be too much of a stretch to conclude that it is the documentation process that ensures that the cyber resilience function is properly run.

In effect, if the organization ever hopes to oversee and refine its control processes, it has to be able to keep track of what those control processes and activities are doing. Consequently, the documentation process has to be planned as part of the overall architecture of the cyber resilience process. That part of the plan articulates a strategy for recording and maintaining all the relevant output of a given cyber resilience control process and/or activity.

Conceptually, the documented realization of a standard model of best practice is at the other end of the application spectrum from the eventual cyber-resilient architectural solution. Because of the requisite tailoring, the solution itself is always organization specific, and generally not transportable to any other organizational setting. However, because it is derived from the standard, the tailored solution DOES represent the profession's most commonly understood best-practice solution to security control assurance in a particular case.

Building the Practical Infrastructure

The cyber resilience concept is based on two essential economic assumptions. First, organizational information is valuable, therefore any loss of information represents a loss of value. Second, the protection of information must be designed and implemented in such a way that the greatest value is obtained for the money invested.

The fact is that you cannot secure everything. Consequently, decisions about what to protect and how to protect it serve as the fundamental criteria in the development of cyber-protection for any organization. Consequently, any given architectural design process has to be capable of arraying a well-defined and functionally similar collection of components into an optimally cost-efficient infrastructure while leaving out functions that are not adequately justified by the value they ensure.

Infrastructure is a generic term that describes a real and applied set of systematic actions that are been put in place to ensure an abstract characteristic such as "cyber resilience." In essence, infrastructures comprise the necessary technology, people, and operational practices to achieve some planned day-to-day operational purpose. The key terms here are "applied" and "real." An infrastructure framework establishes and substantiates practical, rational, and systematic everyday strategic purpose based on a specific investment of real resources. The investment is designed to achieve a defined goal. Each element of the infrastructure has a discrete reason-to-be, and each contributes differently to the ultimate goal, which in this case is a cyber-resilient organization.

Practical infrastructure design involves identifying and documenting all the requisite elements of the eventual solution and then specifying how they interrelate. Any technical, human, or practice behavior that does not achieve the desired organizational purpose would be a waste of money. Therefore, the process of developing the right mix of elements requires the organization to strike a rational balance between the outcomes of the proposed infrastructure component and its associated cost. That process is architectural in the sense that actual structures of related elements are being assembled based on some inherent logic. Therefore, the process of assembling an organizational infrastructure can be compared to, and has the same practical purpose as the architectural considerations that go into designing a complex building.

The end result of the cyber resilience control infrastructure development process is a coherent and fully integrated cyber-resilient architecture. This architecture is never a one-size-fits-all proposition. Instead, the determination of each element of the system must be driven by assessment and the degree of assurance required within a particular setting. And, the rule that underlies that condition is that all situations are different. Therefore, a correctly developed control response is always precisely tailored to correspond to the exact requirements of any given circumstance.

The process requires that there be some kind of standard point of reference to coordinate all security activities that will apply throughout the life cycle, which is the reason why a standard model of expert best practice is required. Cyber resilience control infrastructures are normally based on a commonly accepted, expert, standard model of best practice because there is no science to dictate the correct set of operating procedures.

Instead organizations have to rely on the industry's most up-to-date understanding of the control actions that work. The recommendations embodied in a best-practice standard are meant to serve as the basis for characterizing a wide range of security control functions; in concept, the recommendations of a best-practice standard are not standalone elements. They actually substantiate numerous facets of the security function that taken together constitute aggregate best practice.

Accordingly, as a set, the recommendations of a best-practice standard comprise a complete and tightly integrated control infrastructure solution that when properly configured produces a fully secure situation. In addition, best-practice standards provide a formal mechanism that allows the organization to explicitly specify the rules for control behavior as well as for how ongoing control performance will be assessed. There are eight commonly recognized factors, shown in Figure 8.1, which

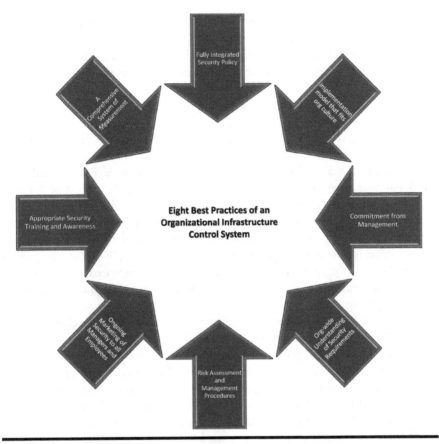

Figure 8.1 Eight best practices of an organizational infrastructure control system.

are considered to be instrumental to the overall success of an organizational infrastructure control system. These are:

1. *The existence of a fully integrated and coordinated, formally documented security policy*: Organizations are governed by their policies. Every policy is operationalized by a targeted set of controls. These controls must be complete, correct, and accomplish the purposes they are set to achieve. The proof that is taking place is confirmed by evidence. Audit is the function that is responsible for collecting, interpreting, and reporting that evidence.
2. *An implementation model that fits with organizational culture*: Human factors are important to the success of almost every element of security. But security behavior is hard to guarantee since it is governed by interpersonal rather than logical rules. Therefore, it is important to ensure that all rules of behavior are defined properly and accompanied by the appropriate motivational approaches. An implementation model that the employees hate will never be successful.

3. *Visible support and commitment from management*: The employees can be expected to resist the imposition of any form of security control. That is because security best practice is generally inconvenient. It makes people do things like enter passwords, when they would rather dive right in. The security people, and even IT administration, don't have the organizational reach or clout to enforce accountability. That can only come from the top. So, the cooperation and commitment of the people in the C-Suite is an integral part of a successful cyber resilience control process.

4. *Organization-wide understanding of security requirements*: Along with the commitment at the top, it is also necessary to make sure that everybody in the organization knows what the rules of behavior are and specifically what their accountabilities are when it comes to security. Understanding implies the need for the design and implementation of formal awareness, training and education programs that are aimed specifically at helping the employees of the organization understand exactly where they fit in and what is required of them.

5. *Formal risk assessment and risk management procedures*: Risk assessment doesn't end when the infrastructure controls are put in place. The risk environment is constantly changing and evolving, and the organization has to evolve its control infrastructure to keep pace. That is the reason why it is essential to have formal risk management embedded in the day-to-day operation as a means of staying on top of the risk picture.

6. *Ongoing marketing of security to all managers and employees*: Security is tiresome, so its necessity has to be explained to the people who do the organization's work. This is an ongoing operational responsibility that is key to sustaining an effective response. That is because people forget over time, and they tend to let inconvenient things slide. So, it is necessary to occasionally remind everybody why they have to do those irritating little security things.

7. *Appropriate training and awareness*: Managers and workers have to be appropriately and specifically trained in their security duties. That is because awareness and training programs make employees aware of the importance of technical and procedural controls. Thus, training and awareness are formal parts of infrastructure control system management. Employees should have scheduled periodic refresher training to assure that they continue to understand and abide by the applicable conditions. Therefore, some form of awareness and training for employees should always be undertaken as part of the overall cyber resilience process.

8. *A comprehensive system of measurement*: Infrastructure controls are monitored and assured by data that is generated from a comprehensive testing and review process. That process involves a carefully planned and targeted set of test, reviews, and audits that are designed to generate substantive data that will allow managers to understand the functioning of the control system at any point in time and for any given situation.

The Detailed Cyber Resilience Control System

A cyber-resilient architecture embeds a set of interrelated or interacting control elements in the organization in order to direct and control how its cyber resilience objectives are achieved. This is accomplished through a process-based cyber resilience control system, which is a network of many interrelated and interconnected activity elements. In that respect, a cyber resilience system is no different than any other complex system. Each activity uses the business' resources to convert some form of input stimulus into a predicted output. Logically, the output of one network action becomes the input of another action.

Therefore, the control activities that take place within a cyber resilience system embody a complex network of input-output relationships, which taken as an integrated whole comprise the single, process-based, cyber resilience assurance solution. To be functionally correct, the solution must specify an exact set of desired outcomes. These specified outcomes allow the organization to determine whether a specified protection goal has been achieved. This includes the specification of the level of validation that is required to authenticate that a given set of results meets the explicit criteria as well as any non-standard, post-task conditions that might be specific requirements of that particular activity.

The purpose of any type of practical architecture is to serve as a foundation for performing a large-scale, real-world business function. And so, in the simplest terms, the architecture of a cyber-resilient process provides the basis for ensuring the long-term security of the organization as a whole. It should be made clear here that architecture is not a management process per se, in the sense that it serves as the cyber resilience architecture. It is not the actual day-to-day, nuts-and-bolts execution of the process. Instead, the infrastructure represents a logical framework within which the work is done. However, to work properly, this structure has to be explicitly stated and its elements have to be explicitly related.

The standard architectural model that provides the basis for developing the applied system must encompass and describe the entire structure of the solution, from top to bottom. Therefore, it is necessary to define every process and practice at a level of detail sufficient to ensure proper operation of the cyber resilience solution. All the details of the elements that are used to categorize that solution must be expressly traceable and derivable from the higher-level elements of the architecture.

Development of a fully defined architecture would be difficult, if not impossible, if it weren't for the fact that standardized process models always contain common features that can be classified into a single category of basic operation. For instance, planning and documentation tasks represent a common set of requirements across most organizations. Yet, these all exhibit pretty much the same entry/task/exit (ETX) practices. Consequently, standard types of activities, like planning or documentation, can be understood and described as a common set of well-defined behaviors. Then, during the actual implementation process that common set can be implemented and interconnected in various ways to achieve the unique purposes of the project.

For instance, planning has a distinct outcome, which is a substantive plan. So, planning always involves the same logical behaviors: information gathering, problem analysis, formulation of a substantive direction, documentation of that direction, development of an attendant monitoring process, and implementation of the plan. Organizations might do some of these individual tasks differently. But the entire set of tasks is still generally executed in that order, and they all are essential components of the process for developing a plan. Thus, there is no need to think through a new set of actions when a planning activity is required. All that is required is to customize the same standard actions to the new situation.

That is the reason why the basic unit of an architectural process model is called a task cell. In practical security implementations, the task cell is also known as a "control." Task cells are unitary functions. That is, each task cell is specified as the means to carry out one specific task, and one task only. The actions the cell defines have logical entry conditions that are required for proper task initiation. These include the inputs from any prior activities at all levels of abstraction. They also must produce an intended outcome from those inputs. The presence of the prescribed outcome serves as the basis for judging whether a task has been carried out properly. There are many task cells in any category of desired outcome in an architecture. The actual tasks themselves, that is, the behaviors that the cell carries out, are defined by describing a specific set of discrete actions that the cell must carry out in order to achieve a commonly understood and well-defined outcome for a given purpose.

These cells or controls are normally tailored to fit the stipulations of a standard model of best practice. That model stipulates the goals that support decisions about how these cells can be interconnected to achieve a higher-level purpose. The value and application of these models will be discussed later in this chapter. But there are other potential sources of association that can be tapped to dictate the possible interrelationships of a task cell. These include: (1) current standard operating procedures within the organization, (2) current or commonly recognized industry methods, and finally (3) any contract stipulations.

Constructing the Process Model for a Particular Application

Once a complete set of standard process cells has been defined, a process model can be constructed for a particular application. This is done by interconnecting the basic set of task cells in various ways to produce a tailored best-practice infrastructure of controls that will satisfy the needs of a given application. The idea is to incorporate into the solution only those behaviors that address the identified issues, and produce the desired outcomes for that particular organizational purpose. In the real world, that involves approaching the solution in three different ways:

1. *A Staged View*: The problem is approached in defined stages.
2. *An Organizational View*: The problem is approached through a model of best practice.
3. *A Control View*: The problem is approached bottom-up through a highly integrated control set applied individually to each requirement, and aggregated into a system.

One of the most important assumptions of this text is that every organization can, and should, implement a cyber resilience process through a formal process, which has been appropriately tailored to fit the requirements of the organization's particular threat environment. The practical mechanism for ensuring this requires the following five steps:

Step One: A standard model of best practice must be selected to define the general form of the infrastructure's architecture, and its constituent controls adopted for tailoring out a complete solution. We have mentioned three popular models in this chapter. However, there are a range of commonly accepted, standard approaches to choose from. None of these is more or less valid in their overall applicability. It simply depends on the threat environment that the adopting organization faces. Obviously, a top-secret organization will adopt the standard model that the government requires. Whereas, it is more likely that a small business will adopt something less demanding, like the COBIT standard. The selection process is governed by the business case, more than it is something based on hard-and-fast rules.

Step Two: Once a standard model has been adopted, a set of behavioral task cells is tailored out of the general activity recommendations. Each behavior must formally operationalize a single control objective within the selected framework.

Step Three: The precise specification of the required input and anticipated output behaviors is done. This allows the organization to monitor and track the behavior of each cell.

Step Four: The specific ETX criteria and expectations must be specified for each cell. This is essential for monitoring ongoing performance.

Step Five: The requisite monitoring is described in an assessment plan. The standard and systematic assessment, measurement, and reporting that are described in that plan are then carried out on a systematic operational basis.

The overall alignment of the tasks in the cyber resilience architecture must be explicitly linked to the recommendations of the standard model of best practice. Yet, because every component in the architecture is unique, due to the tailoring for each situation, the practical solution also reflects the individual requirements of each organizational application. In addition, the resultant cyber resilience architecture cannot be too rigid. The architectural solution has to be capable of modification as a way of reflecting the dynamic changes that are likely to occur in the threat environment over time.

Making Data-Based Decisions about Performance

Besides the need to assess where an organization is in relation to common best practice, using some form of standardized criterion, there is also the requirement that the organization ensure proper and effective decision-making about the operation of their cyber resilience system on an ongoing basis. This implies the need for a formal and systematic measurement and evaluation process that will allow the business to oversee the ongoing execution of the process using objective data. In general, the measurement and evaluation process must address the following practical concerns in order to obtain that data:

■ *Operational Performance of the System*—What are the indicators of proper functioning?
■ *Asset Prioritization*—Which assets are critical versus noncritical? How is this changing?
■ *Risk Acceptance*—What are the risks of recovering versus protecting assets?
■ *Benchmarking*—Are we achieving standard best practice?

In order to answer those questions, the cyber resilience measurement and evaluation process must have a system in place to provide objectively derived assessment data. Specifically, there must be a systematic means to ensure the monitoring and control of the evolution of the cyber resilience process over time. The ongoing understanding of the explicit operational behavior of the formal array of cyber resilience controls is derived from a continuous monitoring process. That process will allow the organization to benchmark control set performance over time. Ideally, management will be able to continuously evaluate the performance of its cyber resilience control infrastructure, and then take substantive action to ensure its effective operation in real time.

The measurement and evaluation process needs to be enforced by explicit, objective assessment measures that will document whether the controls that have been deployed in the cyber resilience system have met the requisite criteria for proper performance. That objective data should provide a basis for management to determine the adequacy of its current operational control set. To achieve that understanding, the measurement and evaluation processes should address the following four operational concerns:

1. Are the cyber resilience controls meeting the business' stated practice specifications?

2. Does the control set continue to achieve generic best-practice recommendations?
3. What are the residual risks of operating the system as it is currently configured?
4. Is the cost of executing the control set justified by value obtained?

The general requirement for systematic execution of the measurement and evaluation process is made operational by a set of well-defined and commonly understood control specifications that can be used to reliably and repeatably benchmark the everyday performance of the organization's real-world control practices. These assessments are done to obtain the data required to confirm control status on a systematic basis. The assessments utilize the outputs of what can be assumed to be proper control performance, which are documented in the standard for best practice that the organization has chosen. The business must perform regular operational assessments using these specifications.

First comes the need to precisely define and document the outcomes that must be obtained in order to indicate proper performance. There are four logical activities that are involved in the evaluation of objective outcomes. The first is the definition of explicit goals. The actions required to achieve those goals must produce objectively quantifiable outputs that can be documented and logged.

Second, the specific behaviors that will be used to characterize those actions and which will be documented for analysis must be described in sufficient detail to ensure that their presence or absence can be unambiguously confirmed.

Third, the specific behaviors that must be present in order to confirm that an activity is being carried out must be specified. That includes their timing and interrelationship. Finally, the prescribed corrective action required, should a behavior not be performed correctly, must be specified and documented. This data must be in objective terms, which can be operationally documented, audited, and verified as correct.

Once the desired outcomes have been itemized in detail, practical, operational measurement and evaluation processes must be designed and deployed to monitor and log those outcomes. Critical success factors that might be documented in this log include:

■ Factors that document the satisfaction of specifically linked policies
■ Factors that document organizational level functions, such as management
■ Factors that document proper execution of the cyber resilience process
■ Factors that document proper execution of an explicitly identified control
■ Factors that document desired technical outcomes

The traditional assessment process is based on the presence of defined and documented outcomes and clear accountabilities. Intangibles such as s strong support/commitment of management, appropriate and effective lines of communication, and consistent measurement practices can also be factored into the assessment process as long as the outcomes can be observed and documented in objective terms.

All these factors belong to the business, and they should always provide unambiguous data that will allow the activity to be substantively measured. Ideally, each factor will be described in terms that will allow the organizational entity that is tasked with the measurement responsibility to be able to determine if, or when, an assessment process is successfully complete.

The practical assessment is performed in precise measurement driven terms. Depending on the outcome of that assessment, there might be a long period of trade-offs and refinement before an eventual decision can be reached about the effectiveness of the cyber resilience function. However, the final documented solution must objectively demonstrate that it addresses the strategic goals and business objectives of the organization.

Implementation Planning

Implementation planning constitutes the next practical step. The outcome is an appropriate set of controls. The security controls are embodied in a comprehensive, practical business architecture directly and verifiably addresses each issue identified in the risk assessment. The design of that architecture is always a creative, conceptual exercise in the sense that its final product is the "blueprint" of the cyber resilience control system that will eventually be put into place. These designs are rarely limited to technical documentation. In case of an overall cyber resilience architecture, the design is generally an overall architectural plan encapsulated in a policy and procedure manual.

All designs exhibit common characteristics. They are complete in the sense that they encompass an entire architectural solution. They are correct in that all elements of the solution, which logically should be present, are there. They are understandable in that they unambiguously communicate the form of the solution. Finally, all the elements of the infrastructure must be traceable to the standard model of best practice that was utilized to shape the architecture.

The design document that encapsulates all these characteristics will establish the quantitative foundations for the measurement function. But in addition, that document must explicitly embody the desired qualitative elements. These elements are subjective rather than objective. Nonetheless, they are useful in judging the success of the design itself. For instance, the stipulation that the system must be "reliable" is an example of a qualitative element.

In addition, the design must provide a clear direction for the integration principle, which is the next step in the process. In that respect, to promote the most efficient interaction, all the technical elements of the proposed solution must be integrated both at their internal interfaces as well as with the elements of the existing operation. Any subjective business process element has the same requirement.

Also, as the name implies, there must be some sort of well-defined and documented planning outcome. All the necessary relationship issues must be identified and resolved in the plan. And, from a resource management standpoint, all risk and control issues have to be coordinated by a planned process. Because a plan provides the explicit direction for the infrastructure development process, it is the essential end product here. In case of a small business, this might be a relatively trivial documentation item, a memo of agreement for instance. However, where the security solution is either very large or complex, there is an implicit element of long-range planning, and the result is always a detailed plan.

In that case, the organization performs the activities associated with a normal strategic planning process. That includes rationalizing the solution against business goals and objectives as well as the formulation of a schedule and a contract. This justification process actively and intentionally aligns the overall form of the cyber resilience response to the organization's existing needs.

Control Integration

Control integration is the largest infrastructure development activity, in terms of the actual time and resources required to execute it. This activity might be just as appropriately termed "realization" or "implementation" because the eventual outcome of the process is the substantive cyber resilience system response. In case of cyber resilience, the general goal of the control system is to ensure authorized access to critical and noncritical assets and make certain that the processing and storage of information within the organization involves a well-defined and effective system of coordinated actions. This type of assurance is typically embodied through five common security attributes, shown in Figure 8.2.

1. *Authentication*: where an individual, an organization, or a computer proves its identity
2. *Authorization*: ensuring access to a specifically constrained collection of assets
3. *Confidentiality*: the ability to maintain the secrecy of assets that must remain private
4. *Integrity*: the assurance of the correctness and accuracy of an asset
5. *Non-repudiation of origin*: All communications are genuine and trustworthy.

The construction of the control infrastructure to assure this passes through eight logical stages. These stages force the business to think through the form of the cyber resilience architecture for that organization. It also ensures that the solution is both optimally resource efficient as well as continuously improving.

Assigning Investment Priorities

The first step in the process logically involves the gathering of all the information necessary to carry out the characterization and prioritization functions that are part of the standard cyber resilience development process. These have already been

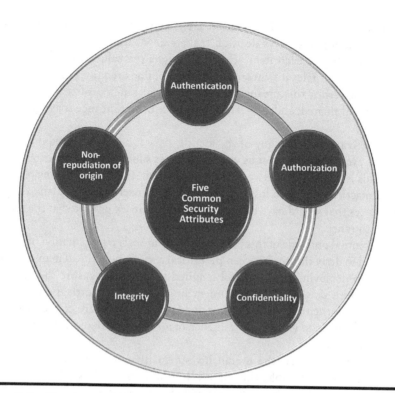

Figure 8.2   Five common security attributes.

extensively discussed in this text. Essentially, that process involves the identification, labeling, and valuation of each and every one of the organization's information assets, and the designation of the critical and noncritical asset baselines. In many respects, this part of the process resembles the asset identification activity that would drive any fiscal accounting or physical inventory process. The individual baselines of critical and noncritical items are differentiated and documented in a formal, organizationally standard ledger. Because that ledger is virtual, it is maintained under the dictates of rigorous, classic configuration management.

Once all of the organization's information assets have been identified and arranged into two coherent baselines, the next step is to determine their priorities. The purpose of this activity is to determine the exact security requirements of each of the individual items in the baseline. It is here that the important decisions are made with respect to the substantive controls that will be assigned to each asset item.

This obviously involves the risk assessment process, which we have discussed, and which is necessary to characterize all the direct threats, vulnerabilities, and weaknesses for each critical baseline item. The outcome of this data gathering process is that all the factors that might impact security for each item are understood and factored into the specific control response.

When the entire set of information assets and their related controls has been identified, the organization's decision makers do a rational assessment to establish the criticality of each item. This step weighs the value of the asset and the implications of all relevant threats against the actions required to address them. In pragmatic terms, this means that the potential impacts of the threat on the business are evaluated, and a level of importance or significance is assigned to the investment.

This activity applies to any of the identified asset item, and it requires consideration of every known threat as well as all other substantive strategic or business issues. This is where the trade-offs actually happen. The working principle here is that the practical impact of loss or harm to a given asset must be balanced against the resources that would have to be invested to ensure it's protected at the requisite level of assurance.

The eventual outcome of this process is a sliding-scale representation of all assets that is used to draw the line between the critical and the noncritical items. The scale ranges from deploying every control required to absolutely ensure the protection of a given asset, down to "the investment in controls isn't worth the value of the asset." This information is then aggregated into a total picture of the investment that must be made to ensure the resilience of all the critical and noncritical assets in the organization.

Since it is likely that the overall investment that is required will be greater than the resources available, this picture will have to be adjusted through negotiations. There will be some assets that will clearly belong in the critical category. Conversely, there will also be assets that will clearly not justify the investment. The art of cyber resilience planning is in navigating the grey area between critical and noncritical. This is where the real value of this part of the cyber resilience infrastructure development process lies.

Decision makers must learn to perform a triage that draws a clear investment policy line. They must prioritize the entire set of asset value and investment concerns in order to ensure that the maximum number of critical functions are guaranteed protected within the resources available. The functioning everyday cyber resilience control infrastructure is the eventual outcome of this process.

Rolling Out the Solution to the Stakeholders

Once the general cyber resilience plan has been developed and approved by the organization's stakeholders, the next step is to customize the real-world set of control practices and procedures. This is the point where the cyber resilience infrastructure is packaged and socialized to the entire organization as a substantive operational process.

The rollout formally establishes the cyber resilience control function itself. It embodies the substance of the cyber resilience process as an everyday reality. It aids the organization's managers in resourcing for the long-haul control deployment and sustainment process. The successful implementation of a functioning standards-based infrastructure depends on the ability to communicate the requirement for a given set of actions to the target organization. Consequently, the rollout of a real-world operational control infrastructure from standards requires a series of focusing steps.

The purpose of these steps is to ensure that the behaviors that will be implemented are fully and completely understood by all stakeholders, and that those stakeholders are capable of fulfilling the purpose and intent of each of the generic recommendations of the standard. To do this, the implementers have to communicate both a high-level understanding of the process, as expressed in the standard, as well as a specific set of control behaviors that implement those recommendations.

The practical realization of the company's cyber resilience process is built around an exact specification of the details of the control infrastructure. In that specification, the requisite control behaviors must be explicitly understood and executed. In order to measure and assure this, the expected behaviors must be executed at a level of unambiguous performance. Therefore, the precise set of actions that must be taken to ensure the protection of a given asset needs to be fully understood by all participants, and a person responsible for each behavior be assigned to every asset.

In order to ensure that, the critical and noncritical asset control behaviors must be fully and completely understood along with the required outcomes. Structurally, the control and organizational dependencies must all be identified and mapped, so that the people who execute the process can understand the various interrelationships of control elements and their interrelationships to the other parts of the solution as a whole.

This mapping will also include a description of each anticipated behavior for each baseline element as well as a value-based justification for their execution. This explanation is important for motivational purposes along with a listing of the controls that have been assigned and the justification for their assignment. The description should also unambiguously specify and document the constituent tasks. In order to ensure proper execution, these are the control tasks that must be performed by each individual and organizational element within that particular process. This description also aids in making resource allocation decisions.

Where standards have been utilized, the requisite practices, conventions, and metrics must be explained and justified along with how compliance will be monitored. The monitoring will include a specific specification of the testing standards and practices as well as the metrics that will be utilized to judge performance. Along with that, the individual tests and their methodologies as well as their metrics will be detailed.

The relevant reviews and audits will also be assigned to the verification of both the implementation process and the operational control infrastructure.

This assignment will stipulate the minimum number of reviews to conduct the audits that will be carried out and when they will be performed.

Finally, the method for problem reporting and corrective action must be itemized. That includes how any identified problems will be reported, and how nonconformances will be tracked and resolved, as well as who will be responsible for those roles. Over a period of time, the organization fine-tunes the controls that it believes are the most effective. Alterations are based on feedback from the stakeholders, and normally this involves a significant period of development time in order to make the process continuously improving.

Operational Measurement

The final concept is operational measurement. In a technical setting, this might be called a metrics program. At its core, the control infrastructure is created to implement a top-level set of assurance requirements. These requirements are derived from the security policies that are generally itemized in the cyber resilience plan and expressed as a formal specification of security requirements. The stakeholder perspectives, such as those of users and managers, are captured in those requirements and are characterized by a set of metrics.

These metrics express the abstract elements of the process in concrete terms. They are essential to ensure common understanding among all participants. For instance, a term like "threats" may have different meanings for managers and technical people. A network security person might characterize a threat as malicious code, intrusions, network interruptions, and denial of service, and would measure it by the things that affect them, such as instances, downtime, or mean-time-to-failure. Managers on the other hand might characterize threats in business terms, such as lost production, cost, and operational interruption time—they would measure it in the terms that are meaningful to them, dollar value, or cost.

Each is an appropriate point of view, and each perspective has a set of measures. The aim is to integrate these differing views and measurements into a single, meaningful, global understanding that meets the security goals. Organizational environment is a critical factor in this process because the rigor and application of the measures selected will vary. For instance, a highly secure government facility requires extensive and rigorous technical security metrics, while a large private organization might be more focused on the measures related to productivity and performance.

Measurement is a fundamental requirement of the cyber resilience control infrastructure implementation and sustainment process. That is because it is difficult to substantively impose any form of managerial control over a virtual asset. Thus, formal measurement programs provide the necessary data that help decision makers evaluate the control infrastructure's ongoing performance as well as maintain the necessary accountability.

As a result, every cyber resilience control infrastructure plan must be accompanied by a comprehensive description of the measurement program. Essentially, the measurement program ensures that confidence in the cyber resilience control infrastructure is maintained through objective data. Thus, the measurement process for any given cyber resilience control infrastructure must be able to provide consistent, data-based monitoring as a means of confirming that specified controls are in place and functioning properly. This is accomplished by performing regular reviews of operational elements at preplanned and mutually agreed on points in time.

Measurement programs are typically founded on a range of standardized metrics. These allow the organization to track and evaluate control performance for any asset in a given situation. Properly set up and maintained, the measurement program will provide critical tracking of the cyber resilience control infrastructure's overall operation and bring any undesired deviations to management's attention. Thus, an explicit plan for conducting validation and verification activities is an essential component of any control infrastructure implementation plan.

There are no universally recognized standard metrics for assessing the performance of security controls. Instead, individual organizations choose the measures that they feel best fit their particular situations. The rule for this is straightforward, whatever metrics are selected, they must be uniformly and consistently applied. In particular, since cyber resilience control operations are oriented toward harm to asset value, there is a need for a uniform definition and of what constitutes a loss of value. This requires the organization to delineate measurable characteristics that it considers to be indicative of harm or loss of value. That process has to be repeated for every potential security concern in order to make the data, produced by the assessment, accurate and meaningful.

The clarification of what constitutes value loss is especially necessary in order to help the measurement program function accurately. The organization can engineer or at least think through its metric requirements by clarifying the potential situations where unsustainable loss might occur. Consequently, it is possible to achieve a consistent and measurement-based description of the control actions using that approach. During the process of thinking this question through, the individual metric items can be identified that are appropriate to any set of assumptions about harm.

Maintaining the Cyber Resilience Control System over Time

The aim of a cyber resilience system is to ensure a dynamic and highly effective response to threats over time. To ensure the ongoing fulfillment of this purpose, the standard cyber resilience control processes must be continuously overseen, maintained, and reformulated if they are found to be deficient. The overall aim is to ensure effective oversight and understanding, reliable long-term sustainment, and disciplined execution of the cyber resilience process. That status has to be maintained in the face of the challenges that arise out of the dynamic environment of organizational threat.

What we have discussed so far are infrastructure development concerns. The organization's cyber resilience architecture is the composite of all the controls that the organization has devised to provide active response and tangible prevention of threats. That control architecture must be maintained in a complete and trustworthy state. That includes the planning and installation processes that implement and assure the operational status of a formally designed and planned cyber resilience architecture. It also establishes and maintains the intentional interrelationships between the elements of the control set.

Nevertheless, standard policies, processes, and methodologies must define and be documented to ensure that the resultant cyber resilience process will continue to be appropriate and effective for a given threat environment. These policies and processes are altered as the threat picture evolves in order to maintain an effective cyber resilience architecture in alignment with the changes in the contextual situation. These policies, processes, and methodologies constitute the tangible elements of the cyber resilience architecture.

Besides creating a tangible architectural response, these policies, processes, and procedures also ensure continuous strategic improvement of the cyber resilience function as the threat picture evolves over time. It should go without saying that the cyber resilience architecture must be continuously maintained in alignment with the threat environment as changes. That involves the development, deployment, and continuous maintenance of the most appropriate and capable set of cyber resilience controls, interrelationships, and technical components.

Thus, the final stage in the cyber resilience process is to devise and implement a set of procedures that will assure the continuous assessment, creation, integration, and optimization of the cyber resilience control architecture. That architecture must be regularly and systematically evolved in order to ensure its currency and long-term effectiveness. To accomplish this, regularly scheduled organization-wide risk assessments must be carried out to ensure the continuous alignment of the cyber resilience architecture with the threat environment, and that architecture must be evaluated to ensure that it remains effective.

Since the field changes constantly, cyber resilience trends should also be assessed in terms of the way they might impact the evolution of the cyber resilience architecture of the organization. Potential control processes should also be evaluated in order to update and refine the cyber resilience architectural strategy. That might include the input from outside sources, such as consultants, in order to ensure that the broadest range of implications and requirements of the cyber resilience architecture are factored into the evolution. The aim is to ensure maximum awareness of the evolving threat environment.

New strategic directions also need to be evaluated as they relate to the ongoing development of the cyber resilience architecture and its substantive controls. From this evaluation, rational decisions can be made about the most effective enterprise-level cyber resilience policies, processes, and methodologies. The aim is to maintain a dynamically effective cyber resilience and controls architecture in the face of inevitable change.

Logically, persistent, long-term control infrastructure sustainment is the mechanism that must be used to control change to the architecture of the cyber resilience control function. Formal control infrastructure sustainment provides two primary advantages. First, it maintains the integrity of all the elements of the control architecture. Second, it allows for the rational evaluation and performance of change to that architecture, as required. A formal sustainment process also gives the organization's decision makers direct input into the evolution of the protection scheme as it changes over time.

Once established, the control infrastructure sustainment process is maintained as an everyday operational process. The goal of the long-term sustainment process is to maintain the infrastructure control set at a defined level of correctness. Of course, that starts from the assumption that a complete and correct control set already exists. So initially, the documentation of the control set must unambiguously demonstrate that the current set of infrastructure controls is both trustworthy and also achieves the stated organizational purpose.

Typically, infrastructure sustainment underwrites the control set's ability to ensure confidence in the continued proper functioning of the cyber resilience function. Sustainment monitors the control set's ability to accurately identify and record problems, analyze those problems, take the appropriate corrective, adaptive, perfective, or preventive action, and confirm the capability of the system to ensure continuing assurance.

Control infrastructure sustainment does this by rationally controlling all changes to the form of the protection scheme. The management level authorized to approve those changes is explicitly defined as part of the overall process of cyber resilience planning. Changes at any level in the basic control structure must be maintained at all levels.

To ensure this, the most current cyber resilience methodologies, processes, and associated documentation must be identified and maintained in a stable state of assurance. Cyber resilience metrics must be developed and collected to support that assurance. These metrics should be used to improve methodology and process efficiency usage. The results of these analyses must be defined and analyzed. Cyber resilience metrics must be standard. They must be used for causal analysis to optimize the ongoing cyber resilience process.

Altogether, to minimize risks, the cyber resilience process needs careful attention to its personnel aspects. Formal teams must be established and coached in how to apply the organizationally standard methodologies and processes. Cyber resilience and control awareness, knowledge of policies, procedures, tools, and standards must be championed and promoted. Therefore, where necessary, cross-organization awareness, training and education processes must be established to communicate the best cyber resilience practices to the employees of the organization. These materials must be appropriate and standard. Their coordination and deployment must be ensured by a formal process.

Chapter Summary

This chapter presents the ideas that underwrite the development of stable, large-scale organizational infrastructures for cyber resilience. It concentrates on the development of a conventional business process that will enable the creation and coordination of a stable, complete, and sustainable cyber resilience response. The goal of such a formal structure is to foster and maintain a complete and consistent environment for the conduct of the cyber resilience process. In order to create a proper infrastructure, the processes and practices that have been developed to fulfill a particular organizational purpose are systematized in such a way that they ensure the long-term, effective implementation of the overall business goal, which in this case is cyber resilience. The cyber-resilient response has to become a part of the overall day-to-day operation of the business in order to have any real impact. That is because every form of effective protection has to be continuously functioning in the everyday business environment.

Thus, cyber resilience must be embedded in the conventional business as an operational process. Most of today's corporate work is carried out in distributed, multilayered, multi-vendor, and even multicultural environments. All that work must be satisfactorily coordinated and controlled in order for the company's stakeholders to trust the outcomes. That coordination and control are often very difficult to achieve given today's complicated business challenges.

Given the inherent complexity in control formulation, the only way to ensure a systematic architectural solution is through a single coherent development strategy. The strategy must ensure a trustworthy, long-term cyber resilience capability, which will address all likely threats, and ensure that all items of critical and noncritical importance are secured appropriate to their potential value. A solution such as this must be systematically designed to merge all requisite behavioral and technical controls into a single, fully coordinated operational process across the organization. Security governance consciously builds an intentional structure of rational elements and interorganizational relationships that will sufficiently ensure the business' information assets.

A security architectural approach centers on the creation of a comprehensive organizational control system and security culture rather than building separate individualized security solutions for each problem. Essentially, the business defines a coherent framework that embodies all the necessary roles, their associated behaviors and practices, and the strategic policies necessary to achieve a single coherent, organization-wide cyber-resilient solution. There are five processes that underwrite the development of a security architectural framework.

The first of these principles is *scope*. In essence, scope addresses protection issues in the light of resource constraints. In the worst case, that might involve leaving a nonessential or low risk resource outside the boundaries of the protection scheme. Knowledge is obtained by formal assessment.

Assessment is the second essential element of the cyber resilience infrastructure development process. That is because assessment is a primary player in any form of risk-based security system development. The assessment function ensures that the organization fully understands the risks inherent in its security environment and the associated set of requirements.

The third element is *best practice*. Best practice is normally embodied in the form of a requisite best-practice reference model, which documents all the necessary controls of a standard governance model. In that sense, the detailed recommendations of a governance model comprise the basis for risk assessment.

Then there is *control definition*. Any one of these frameworks can be used to judge whether a security infrastructure and its constituent activities are complete and correct. Whatever the source, the reference framework should specify and populate a well-defined and detailed organizational best-practice architecture that allows the organization to develop its own explicit control policies, practices, and procedures.

The actual control system is *tailored* top-down to any desired level of detail for any given organizational application. A standard tailoring approach is particularly important in the creation of security infrastructures, since the implementation of the actual security system is always different in its particulars.

In order to be at their most effective, security infrastructure development processes should have both maximum flexibility and standard structure, which might seem like a contradiction. The simple resolution is to create the architecture of the solution top-down from the highest possible level of concept. The largest and most comprehensive view of the solution can then be used as a general classification structure within which effective controls for each of the specific security areas can be addressed. Using this approach, a specific cyber-resilient control architecture can be constructed for any given project at any given level of definition inside the model best-practice framework.

The outcome of the specific tailoring of controls to satisfy each of the general security categories of the standard should be an operational process model, that embodies best practice, and which specifically ensures cyber resilience for a given organizational application.

The behavior of any form of control system has to be documented. Documentation is an absolute ongoing requirement. In many respects, the documentation of control outputs is the absolute bedrock on which oversight and management rest. So, it would not be too much of a stretch to conclude that it is the documentation process that ensures that the cyber resilience function is properly run.

Practical infrastructure design involves identifying and documenting all the requisite elements of the eventual solution and then specifying how they interrelate. Any technical, human, or practice behavior that does not achieve the desired organizational purpose would be a waste of money. Therefore, the process of developing the right mix of elements requires the organization to strike a rational balance between the outcomes of the proposed infrastructure component and its associated cost.

That process is architectural in the sense that actual structures of related elements are being assembled based on some inherent logic. The activities that take place within a cyber resilience system embody a complex network of input-output relationships, which taken together comprise the single process-based cyber resilience assurance solution. To be functionally correct, the architectural model must specify an exact set of exit conditions. That allows the organization to determine whether the specified task has been executed properly in that instance.

The standard architectural model that provides the basis for developing the applied system must encompass and describe the entire structure of the solution, from top to bottom. Therefore, it is necessary to define every process and practice at a level of detail sufficient to ensure proper operation of the cyber resilience solution. All the details of the elements that are used to categorize that solution must be expressly traceable and derivable from the higher-level elements of the architecture.

The basic unit of an architectural process model is the task cell. In practical security implementations, the task cell is also known as a "control." Task cells are unitary functions. That is, each task cell is specified as the means to carry out one specific task, and one task only. The actions the cell defines have logical entry conditions that are required for proper task initiation. These include the inputs from any prior activities at all levels of abstraction. They also must produce an intended outcome from those inputs.

Once a complete set of standard process cells has been defined, a process model can be constructed. This is done by interconnecting the basic set of task cells in various ways to produce a tailored set of best-practice controls that satisfy the needs of a given application. The practical mechanism for ensuring this requires the following five steps:

First: A standard model of best practice must be adopted to define the general form of the architecture, and for tailoring out a comprehensive solution.

Second: Once a standard model has been adopted, a set of task cells is tailored out of the general activity recommendations.

Third: The specific ETX criteria and expectations are specified for each cell.

Fourth: The precise specification of the required input and anticipated output behaviors is done. This allows the organization to monitor and track the behavior of each cell.

Fifth: The requisite monitoring is described in an assessment plan. The standard and systematic assessment, measurement, and reporting that are described in that plan are then carried out on a systematic operational basis.

There must be a systematic means to ensure the monitoring and control of the evolution of the cyber resilience process over time. The ongoing understanding of the explicit operational behavior of the formal array of cyber resilience controls is derived from a continuous monitoring process. That process will allow the organization to benchmark control set performance over time. Ideally, management will be able to continuously evaluate the performance of its cyber resilience control infrastructure, and then take substantive action to ensure its effective operation in real time.

The practical assessment is performed in precise measurement driven terms. Depending on the outcome of that assessment, there might be a long period of trade-offs and refinement before an eventual decision can be reached about the effectiveness of the cyber resilience function. However, the final documented solution must objectively demonstrate that it addresses the strategic goals and business objectives of the organization.

The aim of a cyber resilience system is to ensure a dynamic and highly effective response to threats over time. To ensure the ongoing fulfillment of this purpose, the standard cyber resilience control processes must be continuously overseen, maintained, and reformulated if they are found to be deficient. The overall aim is to ensure effective oversight and understanding, reliable long-term sustainment, and disciplined execution of the cyber resilience process. That status has to be maintained in the face of the challenges that arise out of the dynamic environment of organizational threat.

Once established, the control infrastructure sustainment process is maintained as an everyday operational process. The goal of the long-term sustainment process is to maintain the infrastructure control set at a defined level of correctness. Of course, that starts from the assumption that a complete and correct control set already exists. So initially, the documentation of the control set must unambiguously demonstrate that the current set of infrastructure controls is both trustworthy and also achieves the stated organizational purpose.

Keywords

Acceptable risk: A situation where either the likelihood or impact of an occurrence can be justified

Analysis: An explicit examination to determine the state of a given entity or requirement

Architecture: A designed entity with a consciously designated purpose

Assurance: The set of formal processes utilized to ensure confidence in software and systems

Baseline: A collection of entities related by a similar purpose and time frame

Baseline security: A minimum level of acceptable assurance of proper performance

Benchmarking: Measurement by comparison with standard measures or past performance

Boundaries: A perimeter that incorporates all items that will be secured

Business impact analysis: An assessment of the effect of the occurrence of a given event

Common features: Functionality shared among a number of standard practices

Controls: Activities built into a process designed to ensure a particular purpose

Infrastructure management: Role that oversees and maintains a defined process architecture

Likelihood determination: An assessment of the probability that an event will occur

Monitoring: Specific oversight created by a planned collection and analysis of data

Quantitative management: Decision-making that is supported by empirically derived data

Prioritization: Assignment of importance based on the perceived value

Process: A collection of practices designed to achieve an explicit purpose

Process architecture: The long-term method of organization of the overall information and communication technology (ICT) work

Process specifications: The explicit work rules and requirements of a given operation

Risk: A given threat with a known likelihood and impact

Risk behavior: An action that will produce a defined exposure or risk for the organization

Risk level: The likelihood and impact that are considered acceptable before a response is required

Risk treatment: The specific control response that is planned for a given adverse event

Security controls: Mechanisms designed to ensure proper performance of the process

Security metrics: Quantitative measures of security performance

Security system: Formal collection of controls aimed at mitigating all known threats

Security testing and evaluation: Validation of the secure performance of the control process

Security vulnerabilities: Explicit known weakness that can be exploited by a given threat

Specification: Documentation of explicit requirements of a given system

Stakeholder: Person responsible for a given item or function, generally also the decision maker

Strategic alignment: Assurance that the security actions of the organization directly support their goals

Strategic framework: The generic organizing and control principles that an organization uses to underwrite the management of its information function

Strategic management plan: The prescribed activities to achieve the long-range intentions of the organization

Strategic planning process: A set of rational activities that are undertaken to ensure long-range directions of the organization

System design: Assurance that the architecture of the system meets requisite criteria

Technical (functions): Automated mechanisms designed to ensure secure performance of the process

Testing: Validation of performance of a piece of software or system

Threat: Adversarial action that could produce harm or an undesirable outcome

Vulnerability: A recognized weakness that can be exploited by an identified threat

Index